Artificial Intelligence in Medical Imaging Technology

Euclid Seeram • Vijay Kanade

Artificial Intelligence in Medical Imaging Technology

An Introduction

Springer

Euclid Seeram
Medical Imaging, Faculty of Health
University of Canberra
Burnaby, BC, Canada

Vijay Kanade
Research & Development
AI Researcher, Intellectual Property
Research
Kolhapur, Maharashtra, India

ISBN 978-3-031-64048-3 ISBN 978-3-031-64049-0 (eBook)
https://doi.org/10.1007/978-3-031-64049-0

This Springer imprint is published by the registered company Springer Nature Switzerland AG
The registered company address is: Gewerbestrasse 11, 6330 Cham, Switzerland

This book is dedicated to my two smart, and beautiful granddaughters,
Claire and Charlotte
with love and blessings to you both forever.
(Euclid Seeram)

This book is dedicated to my beloved parents,
late Dr. A.S. Kanade (Lt. Commander) and Prof. Jaya A. Kanade,
whose wisdom, guidance, and love continue to inspire me every day.
With heartfelt gratitude and everlasting affection, this work is a tribute to your unwavering support and enduring legacy.
(Vijay Kanade)

Foreword (*For Euclid Seeram*)

In the rapidly evolving landscape of healthcare, where technology is revolutionizing diagnostics and treatment paradigms, the intersection of Artificial Intelligence (AI) and medical imaging stands as a beacon of innovation and promise. AI is currently finding success in healthcare data integration, workflow optimization, diagnostic assistance, image reconstruction, and treatment planning to name a few. With AI integrating into so many aspects of the radiography department, technologists and therapists are and will be forefront healthcare providers in the training, validating, and implementing of different AI applications. Within this aspect, Dr. Euclid Seeram's work, "Artificial Intelligence for Medical Imaging Technology: An Introduction," emerges as an indispensable guide.

Dr. Seeram's extensive expertise, coupled with his visionary insights, has yielded a comprehensive resource that addresses the pressing need for integrating AI into the domain of medical imaging. As a distinguished authority in both medical imaging and AI, Dr. Seeram navigates the intricate terrain of this field with clarity and precision. Dr. Seeram has published 31 textbooks on various subject matters in medical imaging and has more than 60 papers published in peer-reviewed journals. Dr. Seeram has served as a consultant and guest lecturer to some of the most prestigious imaging societies on an international scale. Dr. Seeram is revered in the medical imaging community for his unwavering expertise and dedication to high standards in quality imaging focused on improving patient care. He is globally renowned as both a distinguished educator and a pioneering researcher. Professionally, we consider ourselves privileged to have collaborated with him on the forefront of the evolving field of AI, particularly within the realm of Computed Tomography (CT). Personally, our CT imaging careers began reading and learning from Dr. Euclid Seeram's "Computed Tomography: Physical Principles, Clinical Applications, and Quality Control." His unwavering dedication to the profession is unmistakably evident in the breadth and depth of his work.

Artificial Intelligence for Medical Imaging Technology: An Introduction is not merely a technical manual; it is a testament to the transformative potential of AI in healthcare. Dr. Seeram elucidates the intricate concepts of AI algorithms, machine learning, and deep learning in a context that demystifies these complex topics for medical imaging technologists and therapists. By providing practical applications and real-world examples, he empowers technologists and therapists to harness the power of AI in enhancing diagnostic accuracy, streamlining workflows, and ultimately, improving patient outcomes.

Artificial intelligence is finding its way into all aspects of life, and medical imaging and radiation sciences are just one area finding ways to utilize it in numerous ways. Crucially, Dr. Seeram recognizes the pivotal role of medical imaging technologists in the integration of AI technologies. As frontline practitioners, technologists are uniquely positioned to leverage AI tools to augment their capabilities, optimize image acquisition protocols, and collaborate effectively with radiologists and clinicians. Dr. Seeram's meticulous scholarship and dedication to advancing the field shines through in every chapter, making this book an invaluable resource for technologists, educators, and healthcare professionals alike offering insights, strategies, and best practices for navigating the AI-driven landscape of medical imaging.

In an era characterized by exponential growth in data complexity and healthcare demands, the need for proficient AI integration in medical imaging has never been more pressing. Dr. Euclid Seeram's work heralds this new era in medical imaging—one defined by collaboration, ingenuity, and the relentless pursuit of excellence. Let us embark on a collective journey toward a future where AI empowers us to deliver the highest standard of care to every patient, in every corner of the globe.

Associate Professor, Weber State University School of Radiologic Sciences
Taylor Ward,
Ogden, UT, USA

Assistant Professor, Arkansas State University Medical Imaging and Radiation Sciences Program
Kendall C. Youngman,
Jonesboro, AR, USA

Foreword (*For Vijay Kanade*)

Medical Imaging industry is already one of the early adopters of the Artificial Intelligence. It is imperative to trace trends and the future of AI deployment in the Medical Imaging domain.

I have known Vijay A. Kanade for several years and I admire his expertise in "Intellectual Property Research." As a versatile academician, his research spans cutting-edge areas such as AI, Bio-inspired computing, IoT, Wireless technology, Neuroscience, Biophysics, Biomedicine, BCI, and so on. With a portfolio of more than 50 research papers published in prestigious journals like Springer, ACM, IEEE, and other renowned Scopus indexed journals and international conferences, Vijay stands at the forefront of innovation and knowledge across multiple disciplines.

His research work has taken him to prestigious universities worldwide, including Massachusetts Institute of Technology (Cambridge, USA), Columbia University (NY, USA), University of California (Santa Barbara, California), Central Michigan University (Michigan, USA), Manchester Metropolitan University (Manchester, UK), National University of Singapore (Singapore), and several others.

I must highlight that Vijay A. Kanade embarked on his research journey at Raman Research Institute, Bangalore, where he involuntarily assumed the role of a "research scholar" for a concise duration of 6 months. During his tenure, he was actively engaged in projects revolving around Brain Computer Interface (BCI). Subsequently, he transitioned into the corporate sector, delving into the realm of "Intellectual Property (IP)," where he served various IP firms for over 8 years. During his tenure as an IP professional, Vijay conducted independent research endeavors, garnering international recognition from universities worldwide, who extended invitations for him to present his work.

Vijay's mission is to advance humanity into as-yet-undiscovered or underexplored scientific realms. He maintains a constant enthusiasm for learning new things and expanding his expertise to diverse subject areas, including medical imaging. His innovative approach to thinking is greatly valued in his research pursuits.

Presently, Vijay serves as an autonomous researcher, dedicating part of his time as an "AI researcher" to corporate firms. Additionally, he remains engaged in addressing IP-related matters whenever required. Owing to his prowess in science and technology, Vijay has been sought after as a seasoned AI expert by media platforms such as Spiceworks.

Vijay belongs to a family of academicians; I am aware that his father, Late Dr. A. S. Kanade was a principal at a renowned college in Goa, a province on the west coast of India and his mother has retired after a long career as a teacher. Thus, spreading knowledge for the greater good is encoded in the DNA of Vijay A. Kanade.

I am glad that Vijay A. Kanade has decided to co-author this book on "Artificial Intelligence in Medical Imaging Technology" which is a timely reference to students studying medical imaging, radiologists, doctors, and medical imaging professionals. This book delves into the basic construction blocks of AI as well as its future and my personal favorites are the chapters that detail out "Principles of Machine Learning" and "Principles of Deep Learning." I am sure that the medical imaging practitioners as well as the researchers in the domain will find this very helpful.

I wish this book all success, which it so richly deserves.

Principal, New Education Society's Arts, Commerce and Science College
Arvind Samb Kulkarni,
Ratnagiri, Maharashtra, India

Intellectual Property Attorney Priyank Gupta,
Pune, Maharashtra, India

Preface

Artificial Intelligence (AI) is everywhere and has entered the realm of Medical Imaging, and applications range from Computer-Aided Detection (CAD), Disease Detection and Classification; Radiomics; Imaging Biobanks; Dose Optimization; Structured Reporting; Workflow; to Interventional Fluoroscopy; Computed Tomography (CT); Image Reconstruction; and Image Processing. The literature is replete with articles and various tutorials describing not only these applications but also the technical aspects of AI including its subfields Machine Learning (ML) and Deep Learning (DL) as a means of introducing important and significant concepts and principles to medical imaging personnel, including radiologists and technologists, biomedical engineering technologists, and administrators. Additionally, the literature has identified the attitudes of these personnel to the notion of AI in applications in Medical Imaging.

A Guest Editorial on *Artificial Intelligence and the Radiographer/Radiological Technologist Profession: A joint statement of the International Society of Radiographers and Radiological Technologists and the European Federation of Radiographer Societies,* published in Radiography 26 (2020) 93–95, states that: *It is of critical importance that radiographers and radiological technologists, as medical imaging and radiotherapy experts, must play an active role in the planning, development, implementation, use and validation of AI applications in medical imaging and radiation therapy, reinforcing the need for the technology to be targeted to the most pressing clinical problems. The optimal integration of AI into medical safety, clinical imaging and radiation therapy can only be achieved through appropriate education of the current and future workforce10 and the active engagement of radiographers and radiologic technologists in AI advancements going forwards.* In a similar vein, the American Society of Radiologic Technologists (ASRT) published a comprehensive white paper in 2020, on *The Artificial Intelligence Era: The Role of Radiologic Technologists and Radiation Therapists.* A summary statement identified several points relating to the preparation for future change, most notably: "Radiologic technologists and radiation therapists should participate in or lead efforts to maintain quality of AI-based devices and to incorporate AI into quality programs, particularly for patient radiation dose.....and should become involved in laying the groundwork for ethical, practical, patient safety, and clinical aspects of AI in their responsibilities and for the betterment of patient care."

The major purpose of this book *Artificial Intelligence for Medical Imaging Technology* is to provide a useful resource to meet the developing educational requirements of the personnel identified above, not only in the United States and Canada, but also in the United Kingdom, Continental Europe, South America, Africa, Asia, Australia, and New Zealand.

The contents in this book are described in 10 Chapters and 4 Appendices as follows:

Chapter 1: Artificial Intelligence in Medical Imaging Technology at a Glance provides a brief orientation to the nature of Artificial Intelligence (AI) and its applications in Medical Imaging Technology. Chapter 2 elaborates on the ideas introduced in Chap. 1 by presenting a comprehensive exploration of Artificial Intelligence (AI) in healthcare, and tracing the historical evolution of AI, the chapter offers insights into machine learning and deep learning. Focusing on clinical decision support systems, and interpretation and validation of AI results for clinical decision-making. Chapter 3 presents a comprehensive overview of machine learning, covering supervised learning techniques like linear regression and decision trees, with applications in medical image analysis. Furthermore, the chapter explores unsupervised learning, reinforcement learning in medical imaging, and outlines critical machine learning components, including data acquisition, model architecture, loss functions, optimization algorithms, and evaluation metrics such as classification, receiver operating characteristic (ROC) analysis, precision-recall curves, and metrics like Intersection over Union (IoU) and Dice Coefficient. Chapter 4 offers a comprehensive exploration of deep learning principles, spanning neural networks like Convolution Neural Networks (CNNs), Recurrent Neural Networks (RNNs), autoencoders, Generative Adversial Networks (GANs), and attention mechanisms. It delves into CNN specifics for image classification, covering convolutional layers, filter sizes, activation maps, and transitions. Crucial topics like dropout, regularization, transfer learning, data augmentation, and CNN interpretability are highlighted, as well as specialized architectures for segmentation, object detection, and image generation, emphasizing their significance in deep learning applications. Chapter 5 delves into the realm of medical image manipulation and comprehension through an exploration of preprocessing techniques, such as noise reduction and image enhancement. Additionally, segmentation methods, feature extraction, and representation within medical images are addressed and image post-processing techniques crucial for obtaining high-quality medical images are reviewed followed by image interpretation and visualization.

Chapter 6 elaborates on several topics introduced in Chap. 1 and outlines the essentials of eight applications of AI in Medical Imaging. These include radiology workflow; computer-aided detection/diagnosis; radiomics; imaging biobanks; disease detection and classification; dose optimization; structured reporting; and image processing. Chapter 7 deals with the elements of image reconstruction algorithms in Computed Tomography (CT), namely, the filtered back projection algorithm and Iterative Reconstruction (IR) algorithms, and their limitations, and finally, a description of how DL algorithms are used in CT image reconstruction. Chapter 8 addresses major elements of Computer-Aided Detection (CADe) and Computer-Aided Diagnosis (CADx). In particular, the main stages of a CAD system are reviewed followed by a brief description of the major elements characteristic of a Deep Learning-based CAD system, highlighting three components are essential in order to provide a useful output to assist with decision making and accurate diagnosis: mass detection; mass segmentation; and mass classification. Chapter 9 delves into the crucial ethical and regulatory aspects surrounding AI in healthcare. It navigates through the regulatory frameworks, such as FDA regulations, UK's Medicines and Healthcare products Regulatory Agency (MHRA), and global agencies that govern AI's integration into healthcare. The chapter explores the concerns regarding privacy and security surrounding medical imaging data. Moreover, it meticulously examines the intricate subjects of bias, fairness, and transparency ingrained within AI algorithms. Lastly, it underscores the imperative of responsible AI use in healthcare, outlining the pivotal considerations for its ethical implementation. Finally, Chap. 10 explores the horizon of possibilities and challenges that define the future of AI in medical imaging. It unfolds a comprehensive exploration of recent breakthroughs in AI for medical imaging, showcasing how cutting-edge technologies are reshaping the diagnostic landscape. From the advent of explainable AI to the promise of federated learning, the chapter navigates through emerging technologies that are steering the trajectory of healthcare innovation. Moreover, the chapter carefully analyzes the potential ramifications of integrating AI into clinical practice and its future in healthcare, providing insights into the opportunities and the hurdles that need to be tackled for responsible incorporation. There are four Appendices related to Chap. 3. The primary aim of these appendices is to offer additional details on specific technical aspects, aiding in the comprehension of the workings of machine learning.

Enjoy, study, and learn from the pages that follow and remember—your patients will benefit from your wisdom.

Burnaby, BC, Canada Euclid Seeram

Kolhapur, Maharashtra, India
Vijay Kanade

Acknowledgments

By Euclid Seeram

An important satisfying task in writing a book of this nature is to thank those medical physicists, biomedical engineers, computer scientists, and manufacturers who have done the original work on the various topics included in this book. First, I would like to thank my co-author Vijay Kanade, BE (CS), who holds a Bachelor of Engineering Degree with a specialization in Computer Science, and who is an Artificial Intelligence (AI) researcher with expertise on a wide range of topics that span the subject matter of AI. I have learned a great deal on AI by just reading and studying his well described and detail chapters on AI included in this book. Furthermore, I have gained additional knowledge on AI from the University of Helsinki's Open Learning course on The Elements of AI. The course features a wide range of topics ranging from Why AI Matters, Probability Fundamentals, and Machine Learning to Neural Networks including Deep Learning Fundamentals.

This book deals with AI in Medical Imaging Technology as outlined in the Preface above, and in this respect, it is indeed a pleasure to express sincere thanks to all of the scientific and clinical experts, whose works have not only been cited but also provided illustrations of various concepts related to AI applications in Medical Imaging.

I am most grateful to Dr. Rob Davidson, PhD, MAppSc (MI), BBus, FASMIRT, Professor of Medical Imaging, University of Canberra, Australia. Thank you for not only assuming one of the roles of supervising my PhD dissertation but more importantly for providing me with the motivation to provide tools to educate our imaging community, and for serving as a reviewer of the initial proposal for this book. Another individual to whom I am grateful is Dr. Patrick Brennan, PhD, of the University of Sydney, Australia, who opened up doors for me to participate in the medical imaging community at the international level. Additionally, I would like to thank the *other reviewers* who provided constructive feedback to our publishers, Springer Nature, in support of this work. They are in alphabetical order, Dr. Anthony Chan, PhD, PEng, CEng, CCE, Program Head, Biomedical Engineering, School of

Health Sciences, British Columbia Institute of Technology (BCIT), and an award-winning Canadian Biomedical Engineer; Dr. Quang Vo, PhD, MSc, BEng, Instructor, School of Health Sciences, BCIT; Dr. Taylor Ward, PhD, Assistant Professor, Weber State University; School of Radiologic Sciences; and Kendall Youngman, MSRS, RT(R)(CT)(MR), Assistant Professor, Medical Imaging & Radiation Sciences Arkansas State University. Thanks for your constructive comments and positive feedback on the need for this book.

I must also acknowledge Yongjun Xu and 31 co-authors for their comprehensive review paper entitled "Artificial intelligence: A powerful paradigm for scientific research" published in *The Innovation 2*, 100,179, 2021, from which I have learned a great deal on the topic of AI, in several domains, such as in Physics and Mathematics and so forth. Another individual to whom I am grateful is Valentina Al Hamouche, MRT(R), MSc, who is the CEO/Founder VCA Education Solutions for Health Professionals based in Toronto, Canada. Valentina has provided me with recurring opportunities to provide Radiographic Imaging Sciences and CT Physics and Technology and other topics such as *AI in Medical Imaging* delivered through live webinars to further educate technologists and students across Canada and internationally as well. Thanks Valentina.

We are grateful to Merry Stuber, Senior Editor, Cell Biology & Biomedical Engineering at Springer, a part of Springer Nature, New York, NY, USA, who did all the hard work in not only reviewing the proposal herself but having it evaluated by several reviewers whose names are mentioned above, that finally led her to accept it for publication. Merry has provided the needed continuous support and encouragement to bring this work to fruition. Additionally, we appreciate the contributions of members of the production team at Springer Nature, including Amrita Unnikrishnan (Ms.), Production Editor, who has worked exceptionally hard during the production of this book, especially in the page-proof stages.

I humbly acknowledge the support from my beautiful family; first my lovely wife, Trish, a warm, smart, caring, and a very special person in my life, thanks babes. Secondly, my caring and very brilliant son David, the best Dad on the planet, to his two most precious daughters, my granddaughters, to whom this book is dedicated. Thanks for your enduring love, support, and encouragement.

By Vijay Kanade

Writing a book of this kind involves a gratifying responsibility of acknowledging the contributions of various professionals whose original work forms the backbone of its content. Firstly, I extend my gratitude to my co-author Euclid Seeram, an esteemed figure in academia, radiology technology, and a well-known author in Medical Imaging. Collaborating with someone of his caliber has been an honor, and I've gleaned valuable insights from his expertise, particularly in emphasizing the importance of paying attention to finer details and also tailoring the content to suit medical professionals, notably radiologists and imaging technologists.

I also want to express sincere appreciation to the multitude of researchers, authors, and experts whose work, referenced and exemplified throughout this book, illuminates the role of AI in medical imaging.

Special thanks are owed to Merry Stuber, Senior Editor at Springer, who diligently evaluated and facilitated the publication process, supported by a dedicated team at Springer Nature. Merry's continuous support was indispensable in bringing this work to reality. We also wish to acknowledge the valuable contributions and exceptional dedication of the production team at Springer Nature, particularly Amrita Unnikrishnan, Production Editor, who worked tirelessly during the book's production.

Lastly, I owe a debt of gratitude to my family for their resolute backing: my mother, Jaya Kanade, who has been a constant source of strength and encouragement; my late father, Dr. A. S. Kanade, whose remarkable professional, academic, and research accomplishments consistently drive me to push the boundaries and strive for greatness, even in his absence; and my wife, Kalyani, a computer engineer and a talented interior and graphic designer, whose artistic abilities brought the illustrations to life. I deeply appreciate their steadfast love and support.

Contents

Chapter 1
Artificial Intelligence in Medical Imaging Technology at a Glance

1.1 Introduction

Medical imaging technology has evolved from film-based imaging modalities to digital imaging modalities. These modalities include digital radiography, digital fluoroscopy, digital mammography, computed tomography (CT), diagnostic medical ultrasound, nuclear medicine, single-photon emission computed tomography, position emission tomography, and magnetic resonance imaging (MRI). These technologies produce both anatomical and functional images that can also be fused together to produce what is referred to as fused images. These images are viewed and interpreted by radiologists (who are specially trained in the art and science of image interpretation) to provide a diagnosis of the patient's medical/health condition. Since this task requires using a large knowledge base to extract various features from images as well as interpret these features, medical imaging is hence well-positioned for AI applications.

Furthermore, medical imaging involves additional tasks and processes other than image interpretation. Figure 1.1 provides examples of other areas of imaging "with potential for AI assistance and enhancement" [1]. These include processes before image acquisition, during image acquisition, following image acquisition, image interpretation, and following image interpretation.

With the above in mind, Potocnik et al. [2] point out that "radiographers, clinicians, and radiation therapists should acquaint themselves with the new technology and demonstrate readiness to actively contribute toward development, implementation, and adoption of AI-enabled solutions."

The purpose of this chapter is to introduce a broad overview of AI. First, AI and its subsets, machine learning (ML) and deep learning (DL), will be defined followed by a brief description of the major features of AI including how AI works. Furthermore, specific applications of AI in medical imaging will be reviewed, and a

E. Seeram, V. Kanade, *Artificial Intelligence in Medical Imaging Technology*,
https://doi.org/10.1007/978-3-031-64049-0_1

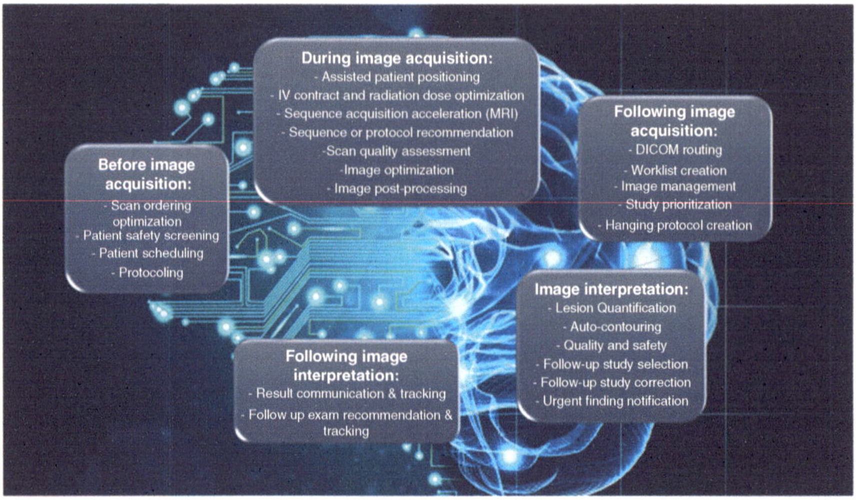

Fig. 1.1 Overview of the process cycle in radiology and major tasks with potential for AI assistance and optimization. (From Pierre K, Haneberg AG, Kwak S, Peters KR, Hochhegger B, Sananmuang T, Tunlayadechanont P, Tighe PJ, Mancuso A, Forghani R (2023). Applications of Artificial Intelligence in the Radiology Roundtrip: Process Streamlining, Workflow Optimization, and Beyond. Seminars in Roentgenology, 58 (2), 158–169. (Reprinted under the CC BY-NC-ND license (http://creativecommons.org/licenses/by-nc-nd/4.0/))

case study of how AI is used in CT image reconstruction will be outlined. Finally, the chapter concludes with a glimpse of the ethics of AI in medical imaging.

1.2 A Little Bit of History

The beginnings and history of AI involve several milestones and breakthroughs, as outlined by Najjar [3] dating back to 1956 when it was formally introduced at the Dartmouth Conference [4]. These milestones and breakthroughs are sketched out in Fig. 1.2 and described by Nijjar [3]. Notable events include the creation of ELIZA, the first chatbot developed at the Massachusetts Institute of Technology (MIT), a decline in AI interest and research in the 1970s, development of neural networks in 1986, chess gaming in which International Business Machine's (IBM's) Deep Blue defeated chess champion Garry Kasparov in 1997, introduction of Google's self-driving car in 2009, and so on to the currently popular Generative Pre-Trained Transformer-4 (GPT-4), an AI model based on deep learning algorithms, which accepts image and text inputs and emits text outputs that "while less capable than humans in many real-world scenarios exhibit human-level performance on various professional and academic benchmarks" [5], which is introduced in 2023.

Today, AI has become commonplace in many areas of society, including medical imaging technologies, and applications continue to increase almost exponentially

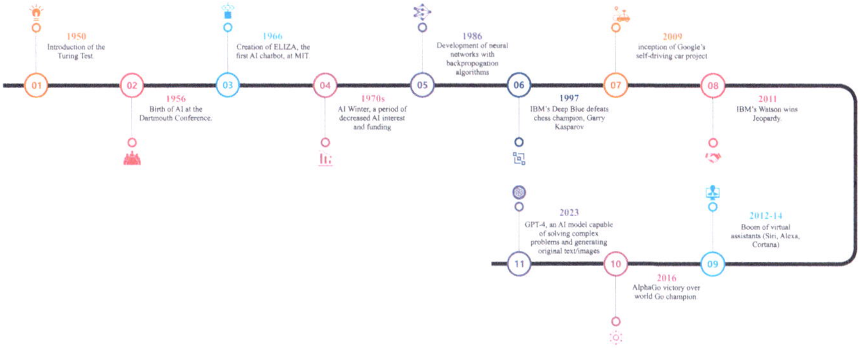

Fig. 1.2 Significant milestones in the evolution of artificial intelligence. (From Najjar, R. (2023) Redefining Radiology: A Review of Artificial Intelligence Integration in Medical Imaging. Diagnostics, 13, 2760. https://doi.org/10.3390/diagnostics13172760. Reprinted under the Creative Commons Attribution (CC BY) license (https://creativecommons.org/licenses/by/4.0/))

all over the world. For a detailed coverage of AI and its "impact on human societies," the interested reader should refer to a textbook by Dr. Christo El Morr [6].

1.3 Definitions

AI is a branch of computer science. It is an extensive subject and is based on principles and concepts related to mathematics, logic, philosophy, cognitive science, biology, and evolution. Key elements of AI include computer vision and scene recognition, robotics and sensory systems, natural language processing, neural computing, expert systems, intelligent computer-aided instruction, and neural networks. Furthermore, these elements provide AI applications in everyday life, including smart phones, smart cars, smart homes, email, banking, social networking, online shopping, sports, and the Internet of Things (IoT), to mention only a few. More importantly and within the context of this book, AI is now used in medicine and healthcare and medical imaging technologies, including computed tomography (CT) [7–10].

1.3.1 Artificial Intelligence

The literature offers a vast number of definitions of AI and its subfields, ML and DL, as illustrated in Fig. 1.3, in which ML is a subfield of AI and DL is a subfield of ML. Each of these terms is defined briefly by Pesapane et al. [11] For example, AI is defined as "a device that mimics cognitive function" (referring to human thinking and problem-solving), ML refers to "algorithms that improve as they are exposed to more data," and DL is "an artificial neural network structured in multiple

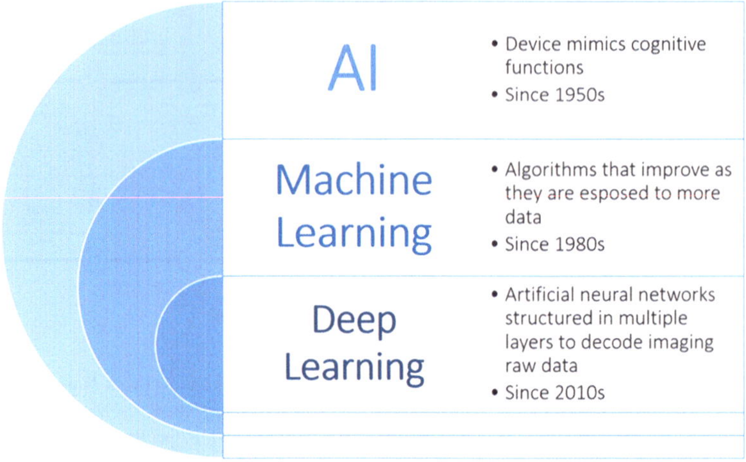

Fig. 1.3 AI and its subfields, ML and DL. While ML is a subfield of AI, DL is a subfield of ML. (From Pesapane, F., Codari, M. & Sardanelli, F. (1918). Artificial intelligence in medical imaging: threat or opportunity? Radiologists again at the forefront of innovation in medicine. Eur Radiol Exp 2, 35. https://doi.org/10.1186/s41747-018-0061-6. Reprinted under the Creative Commons Attribution 4.0 International License (http://creativecommons.org/licenses/by/4.0/))

layers to decode imaging raw data." The word algorithm refers to a finite set of rules for solving a problem. These rules can be coded for computer processing.

A more detailed statement of the meaning of AI is offered by Nijjar [3] as "the replication of human intelligence in machines that are programmed to emulate human cognition and actions, encompassing learning, problem-solving, reasoning, and perception. AI can be classified into two major types: narrow AI, which is designed for specific tasks (e.g., facial recognition or voice commands), and general AI, which mimics a broader spectrum of human intellect. AI aims to develop systems with autonomous intelligent functionality, capable of problem-solving, decision-making, and performing tasks typically requiring human intelligence."

1.3.2 Expansion of AI

There are three fundamental elements as shown in Fig. 1.4 driving the expansion of AI in an enormous way. These include scientific breakthroughs, greater computing power, and more data. Furthermore, this expansion has been driven mostly by the application of machine learning and deep learning [12].

While scientific breakthroughs refer to discoveries that change knowledge and have a significant impact on science and technology as well as an impact on society as a whole, computing power depends on the number of processors (more processors mean more power) and the processor's speed (faster speed means greater computing power). Additionally, using graphics processing units (GPUs) for parallel

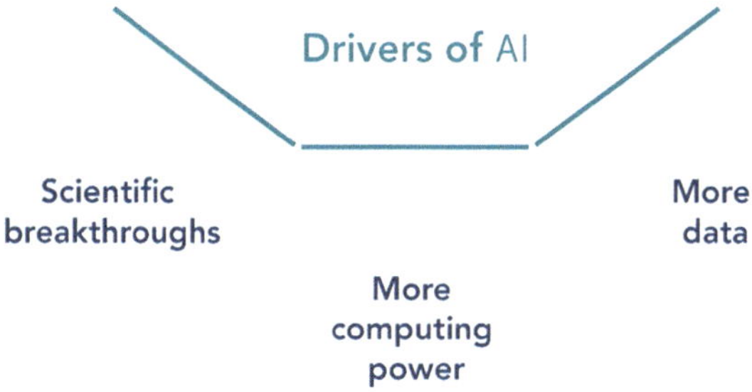

Fig. 1.4 Three fundamental elements driving the expansion of AI. See text for further explanation. (From Sheikh, H., Prins, C., Schrijvers, E. (2023). Artificial Intelligence: Definition and Background. In: Mission AI. Research for Policy. Springer, Cham. https://doi.org/10.1007/978-3-031-21448-6_2 Reprinted under the Creative Commons Attribution 4.0 International License (http://creativecommons.org/licenses/by/4.0/))

processing increases the computing power. Medical imaging has experienced scientific breakthroughs such as dose reduction methodologies, CT image reconstruction algorithms, and more recently photon counting CT technology and now uses GPUs as part of its instrumentation architecture. Furthermore, medical imaging is now considered a big data profession, since it generates very large amounts of data. Big data is characterized by the four Vs, namely, volume, velocity, variety, and veracity. While volume refers to the amount of data, velocity addresses the speed with which new data is generated and propagated. Variety, on the other hand, refers to the fact that the data can be structured, unstructured, and semi-structured. Finally, veracity of big data examines whether the data is trustworthy.

AI is explained in a little more detail in Chap. 2.

1.3.3 Machine Learning

As shown in Fig. 1.3, ML is a subfield of AI. The literature includes several definitions of ML. For example, two such definitions include the following:

1. ML is "the use and development of computer systems that are able to learn and adapt without following explicit instructions, by using algorithms and statistical models to analyze and draw inferences from patterns in data" [13].
2. ML "is an AI technique that teaches computers to learn from experience. Machine learning algorithms use computational methods to 'learn' information directly from data without relying on a predetermined equation as a model. The algorithms adaptively improve their performance as the number of samples available for learning increases. Deep learning is a specialized form of machine learning" [14].

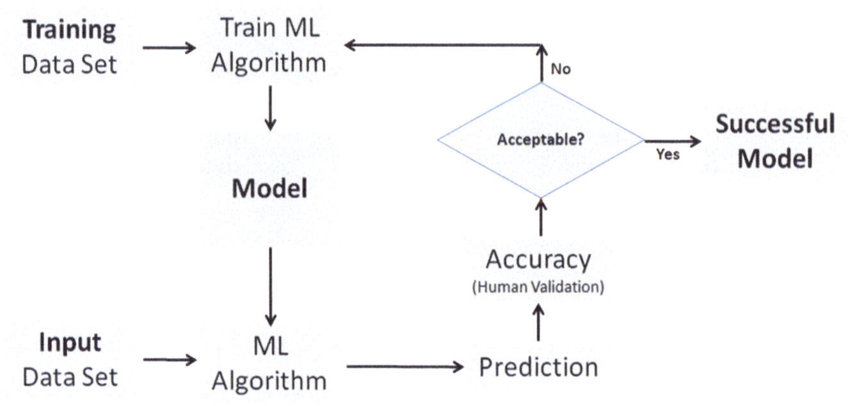

Fig. 1.5 The fundamental steps involved in ML. See text for further explanation. (From Seeram E (2023) X-Ray Imaging Systems for Biomedical Engineering Technology. Springer Nature Switzerland AG, 2023. Reproduced by permission)

The major characteristics of ML are its algorithms and learning categories. The learning categories include supervised learning, unsupervised learning, semi-supervised learning, and reinforcement learning, while the algorithms are neural networks, linear regression, logistic regression, clustering, decision trees, and random forests [14]. These topics are not within the scope of this chapter; however, they will be described in Chaps. 3 and 4.

Figure 1.5 illustrates the fundamental steps involved in ML. First, training data is needed to train the ML algorithm and the outcome of training is called a model. New data is then entered into the model, and the algorithm will subsequently make predictions on the basis of observation, experience, or scientific reason. Predictions are then validated by humans for accuracy. If the output result is acceptable, then the ML model is successful. If the output result is not acceptable, then the process repeats itself until the desired result is obtained. ML will be described in more detail in Chap. 3.

1.3.4 Deep Learning

DL is a subfield of ML (Fig. 1.3) and has been described as being more complex than ML. Numerous definitions are offered in the literature [15–17]. Two such definitions, for example, are the following:

1. DL "is essentially a neural network with three or more layers. These neural networks attempt to simulate the behavior of the human brain—albeit far from matching its ability—allowing it to 'learn' from large amounts of data. While a neural network with a single layer can still make approximate predictions, additional hidden layers can help to optimize and refine for accuracy" [15].

2. DL "utilizes multilayered artificial neural networks (aptly coined 'deep'), enabling DL algorithms to model and understand complex data patterns. These algorithms are especially effective for tasks where manual feature extraction proves challenging, such as image or speech recognition. It is noteworthy that these feature layers are data-learned, not human-engineered, embodying an architecture inspired by the human brain and have proven successful in visual recognition tasks, even outperforming human performance" [16, 17].

The above definitions identify two key characteristics, namely, neural network and learning from data. DL is more complex that ML. Learning from data involves different kinds of models employed in DL. For example, the model used for image classification is not the same as the model used for speech classification [18]. Other topics related to DL include neural networks and deep learning architectures such as convolutional neural networks (CNNs), generative adversarial networks (GANs), and architectures for segmentation, object detection, and image generation. The dominant framework very suitable for use in medical imaging [where pixel data (digital images) are used] is the CNN, the basic structure of which is illustrated in Fig. 1.6, which shows three groups of layers, namely, convolutional layer, pooling layer, and fully connected layer, as well as activation functions. Various terminologies are also introduced such as feature maps, pooled feature maps, vectorization, factorized feature maps, fully connected layer, and output. CNNs are extremely complex and have been applied exclusively in medical imaging since these models use pixel data in images as input into the DL engine [15]. CNNs have been applied exclusively in CT image reconstruction to reduce image noise for low-dose CT

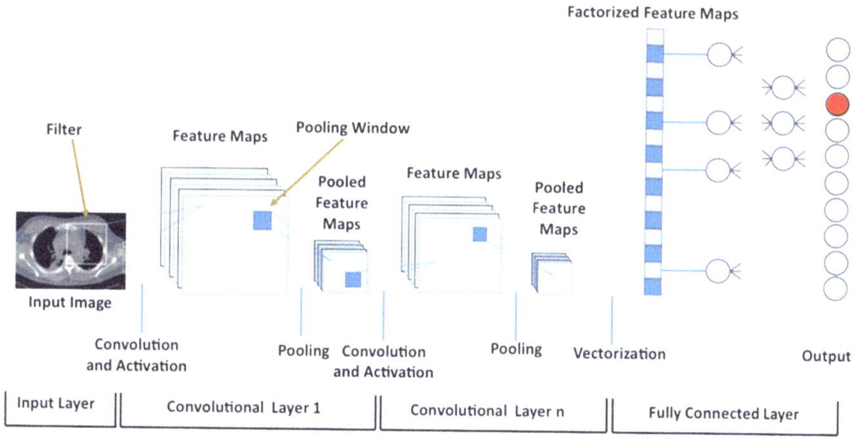

Fig. 1.6 The general structure of a convolutional neural network (CNN). See Chap. 4 for further explanation. (From Pesapane, F., Codari, M. & Sardanelli, F. (1918). Artificial intelligence in medical imaging: threat or opportunity? Radiologists again at the forefront of innovation in medicine. Eur Radiol Exp 2, 35. https://doi.org/10.1186/s41747-018-0061-6. Reprinted under Creative Commons Attribution 4.0 International License (http://creativecommons.org/licenses/by/4.0/))

(LDCT). These algorithms have been referred to as deep learning image reconstruction (DLIR) algorithms [19, 20].

Major elements of a neural network and an artificial neural network (ANN) are neurons, similar to those in the human brain. ANN is characterized by an assembly of hundreds of interconnected artificial neurons (basic computing units) and trained by an algorithm "where pairs of input signals and desired output decisions are presented, mimicking the condition where the brain relies on external sensory stimuli to learn to achieve specific tasks" [21]. This model is a mathematical model of a biological neuron called a perceptron [18]. DL will be described in more detail in Chap. 4.

1.4 Which Subfield of AI to Use in Medical Imaging?

A scheme to assist in making a decision about the use of ML or DL in a facility is shown in Fig. 1.7. Since digital medical imaging operations generate large digital datasets (vast amounts of digital data) and use good computer hardware [such as the graphics processing units (GPUs)], DL algorithms are more suitable and reliable for use in medical imaging AI tasks [15–17, 21].

1.5 Imaging Modalities Using AI Applications

In 2021, Rezazade Mehrizi, van Ooijen, and Homan [22] conducted a systematic review referred to as technography (the study of technological developments in a domain of application) to explore the imaging modalities to which AI was applied. As shown in Fig. 1.8, most of the AI applications target modalities such as CT, MRI, and X-ray imaging modalities. Very few applications were in ultrasound (9%) and in mammography (8%). Most of the applications (95%) work with only one single modality. Only eight applications (3%) work with both CT and MRI modalities. The

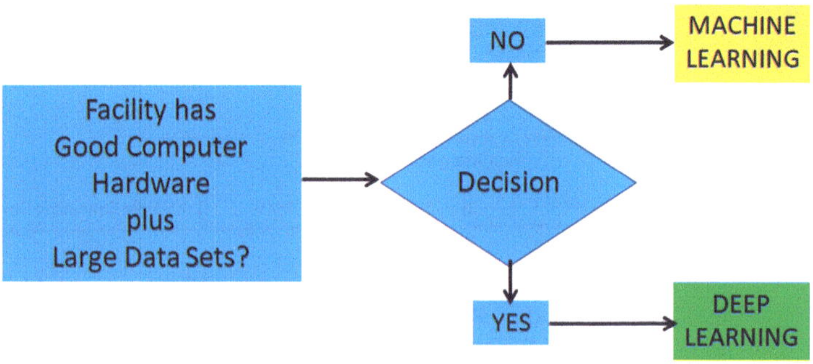

Fig. 1.7 A scheme to assist in making a decision about the use of ML or DL in a facility

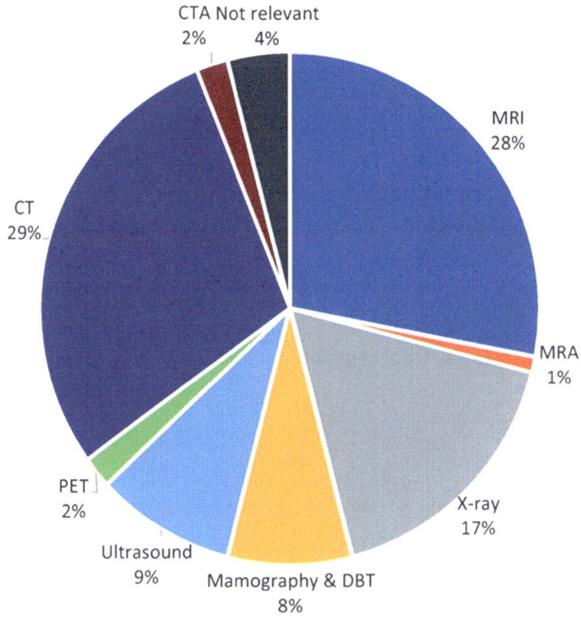

Fig. 1.8 The relative share of applications based on their targeted modalities. (From Rezazade Mehrizi, M.H., van Ooijen, P. & Homan, M. Applications of artificial intelligence (AI) in diagnostic radiology: a technography study. Eur Radiol 31, 1805–1811 (2021). https://doi.org/10.1007/s00330-020-07230-9. Reprinted under the Creative Commons Attribution 4.0 International License (http://creativecommons.org/licenses/by/4.0/))

study concluded that "the researchers, developers, and medical practitioners need to trace and critically evaluate the technological developments, detect potential biases in the way these applications are developed, and identify further opportunities of AI applications" [23].

1.6 How AI Works in a Nutshell

The literature is rich on articles and books describing how AI works, and these descriptions range from simple to complex. This chapter will use the explanation provided by Kanade [23], an AI researcher and co-author of this book, the basic steps of which are illustrated in Fig. 1.9 as follows:

1. The AI system accepts data input in the form of speech, text, image, etc.
2. The system then processes data by applying various rules and algorithms, interpreting, predicting, and acting on the input data.
3. Upon processing, the system provides an outcome, i.e., success or failure, on data input.

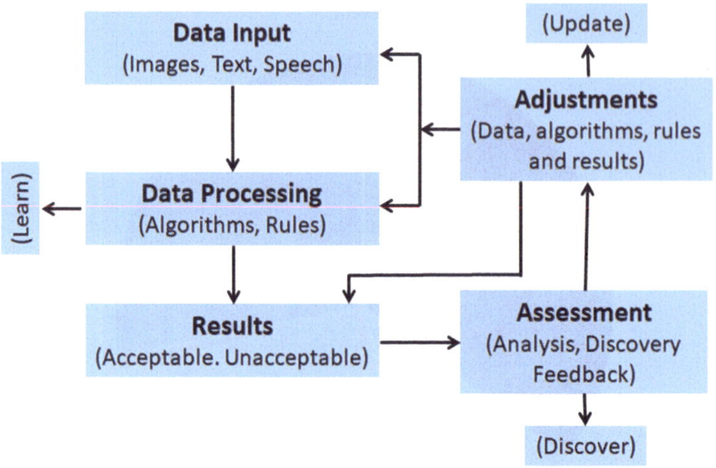

Fig. 1.9 The basic steps of how AI works as outlined by Kanade [23]. See text for further explanation

4. The result is then assessed through analysis, discovery, and feedback.
5. Lastly, the system uses its assessments to adjust input data, rules and algorithms, and target outcomes. This loop continues until the desired result is achieved [22].

1.7 Applications of AI in Medical Imaging

In medical imaging, images are usually interpreted by radiologists who specialize during their education and training in *anomaly detection* and other visual performance strategies such as segmentation and image classification that allow them to make a diagnosis. Anomaly detection [24] is an entire field of study (and not within the scope of this book). One of its central features is "recognizing abnormal inputs based on the seen examples of normal data" [25]. Common anomalies in the brain include glioblastomas, multiple sclerosis (MS), and cerebral infarction (CI) [26]. This knowledge allows radiologists to perform lesion detection and characterization tasks when interpreting images to provide a diagnosis. These tasks are well suited to using AI as a tool in image interpretation. Furthermore, pre-processing techniques such as noise reduction, image enhancement, image segmentation methods for extracting regions of interest, feature extraction and representation of medical images, and medical image classification will be explained in detail in Chap. 5.

An early paper published by the European Society of Radiology (ESR) identifying several applications of AI in radiology [20]. These applications range from computer-aided detection; disease detection and classification in mammography, chest radiography, ultrasound and CT, and other tumor imaging techniques; radiomics; imaging biobanks; dose optimization; and structured reporting. It is not within the scope of this chapter to describe all of these applications, and therefore the reader should refer to the ESR paper [20] for more information. The following points of each of these applications [20] are highlighted below and reproduced under the terms of the Creative Commons Attribution 4.0 International License (http://creativecommons.org/licenses/by/4.0/):

- *Computer-aided detection (CAD)*—CAD makes diagnoses for which it has been specifically trained and bases its performance on a training dataset and a rigid scheme of recognition that can only be improved if more datasets are given to the CAD algorithm.
- *Disease detection and classification*—Major applications include breast cancer detection in mammography and MRI, interpretation of chest radiographic images in chest CT, and other tumor imaging.
- *Radiomics*—Extraction of features from diagnostic images, the final product of which is a quantitative feature/parameter, which is measurable and mineable from images.
- *Imaging biobanks*—Increasing storage of computer-generated data in radiology. Increasing data storage in picture archiving and communication systems (PACS) is one such idea.
- *Dose optimization*—AI can be an optimizing tool for assisting the technologist and radiologist in choosing a personalized patient's protocol, in tracking the patient's dose parameters, and in providing an estimate of the radiation risks associated with cumulative dose and the patient's susceptibility (age and other clinical parameters).
- *Structured reporting*—AI can aid the reporting workflow and help the linking between words, images, and quantitative data and finally suggest the most probable diagnosis.
- *Daily workflow*—AI tools can also impact the daily workflow by filtering exam priority based on appropriateness criteria. ESR has implemented the "ESR iGuide," a clinical decision support system (CDS) that assists referring physicians to choose the most appropriate imaging procedure based on the level of evidence for appropriateness, and the level of emergency AI coupled with CDS can improve the decision process and thereby optimize clinical and radiological workflow.

Another notable application of AI in medical imaging is that of *digital image processing* operations such as image segmentation, feature extraction, and image classification [27]. Image processing and analysis will be described further in Chap. 5.

1.8 AI-Based Image Reconstruction in Computed Tomography

The goal of the use of AI-based algorithms in CT image reconstruction is to enable LDCT imaging while maintaining good image quality and less dose to patients compared with previous reconstruction algorithms such as the filtered back projection algorithm and iterative reconstruction algorithms. Two such CT scanners were approved by the US Food and Drug Administration (FDA) in 2019 including the TrueFidelity from General Electric (GE) Healthcare [28] and the Advanced Intelligent Clear-IQ Engine (AiCE) from Canon Medical Systems [29]. Both of these scanners use DL algorithms to accomplish image reconstruction in low dose CT imaging. The goal of the DL algorithm in both scanners is to reconstruct images using both low-dose CT (LDCT) data as well as high-dose raw data (sinogram) to generate what is referred to as the ground truth image. The details of how these two scanners use DL will be elaborated in Chap. 8.

1.9 Ethics of AI in Medical Imaging

The application of AI principles in healthcare and medical imaging has prompted the issue of ethical considerations of such use in these vocations. This topic is becoming increasingly commonplace in all areas where AI is being currently being used. Ethics is a complex subject matter and not within the scope of this book; however, it is important that a glimpse into the nature of ethics is noteworthy in providing guiding principles for observing ethics in clinical practice. One brief description of the meaning of the term ethics is as follows:

> The term ethics may refer to the philosophical study of the concepts of moral right and wrong and moral good and bad, to any philosophical theory of what is morally right and wrong; or morally good and bad; and to any system or code of moral rules, principles, or values. The last may be associated with particular religions, cultures, professions, or virtually any other group that is at least partly characterized by its moral outlook [30].

The increasing attention of the ethical use of AI medical imaging has prompted radiological organizations to address ethical guidelines and statements on AI use in radiology [31, 32].

One notable example is the joint European and North American multi-society statement. This paper [32] discusses topics such as the ethics of data, ethics of algorithms and trained models, and the ethics of practice. According to the statement, ethical use of AI should be appropriately transparent and highly dependable, curtail bias in decision-making, ensure benefits and harms are distributed among stakeholders in a just manner, ensure responsibility and accountability remain with human designers or operators, minimize harm, and promote well-being. Ethics and regulatory considerations will be discussed in Chap. 8.

1.10 The Radiological Technologist/Radiographer in AI

The majority of papers addressing the issue of using AI in medical imaging are specifically focused on the application of AI in anomaly detection [15, 6, 20, 21, 24–26] and image reconstruction [28, 29] in CT, both of which are in the domain of radiologists and medical physicists [33]. Not surprisingly, however, the literature offers more and more papers on the role of technologists/radiographers in the AI era. For example, the American Society of Radiologic Technologists (ASRT) [34], the International Society of Radiographers and Radiological Technologists (ISRRT), the Society and College of Radiographers in the United Kingdom, and the European Federation of Radiographer Societies (EFRS) [35], and several papers [36–39], have examined this issue. In general, the discussion and recommendations have provided outcomes that stress the need for technologists and radiographers to become educated in AI, ML, and DL and their applications in medical imaging technologies, in order to communicate effectively with radiologists, medical physicists, and manufacturers, and more importantly to assume leadership roles in their profession.

1.11 Deep Learning Implementation and Distinguishing Characteristics

In summary, DL has found widespread applications in medical imaging as described by an example of implementation, and distinguishing characteristics are illustrated in Fig. 1.10. In a recent paper entitled "A holistic overview of deep learning approach

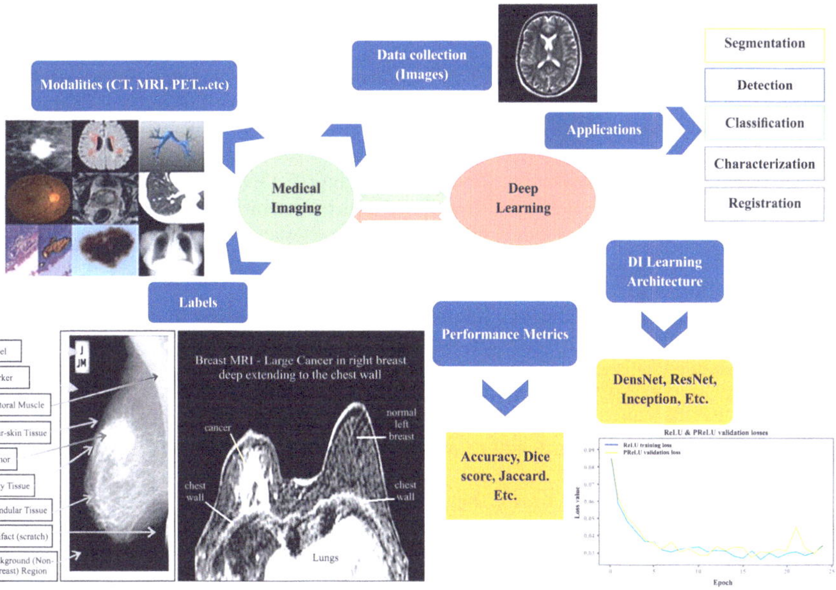

Fig. 1.10 An overview of deep learning implementation and a few distinguishing features for medical imaging application. (From Yousef, R., Gupta, G., Yousef, N. et al. (2022). A holistic overview of deep learning approach in medical imaging. Multimedia Systems; 28, 881–914. https://doi.org/10.1007/s00530-021-00884-5. Reproduced by permission)

in medical imaging" published in *Multimedia Systems*, Yousef et al. [40] "present several applications of deep learning (e.g., segmentation, classification, detection, etc.) which are commonly used for clinical purposes for different anatomical site, and we also present the main key terms for DL attributes like basic architecture, data augmentation, transfer learning, and feature selection methods. Medical images as inputs to deep learning architectures will be the mainstream in the coming years, and novel DL techniques are predicted to be the core of medical images analysis" [40]. These characteristics and many important and significant specific technical AI topics will be described in Chaps. 2, 3, 4 and 5.

1.12 Future Trends and Challenges

Future trends and challenges are part of the evolution of technologies in medical imaging and AI. These will be discussed in Chap. 10.

References

1. Pierre K, Haneberg AG, Kwak S, Peters KR, Hochhegger B, Sananmuang T, Tunlayadechanont P, Tighe PJ, Mancuso A, Forghani R (2023). Applications of Artificial Intelligence in the Radiology Roundtrip: Process Streamlining, Workflow Optimization, and Beyond, Seminars in Roentgenology, 58 (2), 158–169,
2. Potocnik Jaka, Foley S, Thomas E (2023). Current and potential applications of artificial intelligence in medical imaging practice: A narrative review. Journal of Medical Imaging and Radiation Sciences 54–61
3. Najjar, R. (2023) Redefining Radiology: A Review of Artificial Intelligence Integration in Medical Imaging. Diagnostics, 13, 2760. https://doi.org/10.3390/diagnostics13172760
4. Buchanan, B.; Shortliffe, E. (1984). Rule-Based Expert Systems: The MYCIN Experiments of the Stanford Heuristic Programming Project; Addison-Wesley Longman Publishing Co., Inc.: Boston, MA, USA.
5. GPT-4 (2023). https://openai.com/research/gpt-4
6. El Morr, C. (2023). AI and Society: Tensions and Opportunities. New York, CRC Press.
7. Do S, Song KD, Chung JW (2020). Basics of Deep Learning: A Radiologist's Guide to Understanding Published Radiology Articles on Deep Learning. Korean J Radiol; 21(1):33–41. https://doi.org/10.3348/kjr.2019.0312. PMID: 31920027; PMCID: PMC6960318.
8. Santosh KC, Antani S, Guru DS, Dey N (2020) Medical Imaging: Artificial Intelligence, Image Recognition, and Machine Learning Techniques. Boca Raton, FL; CRC Press-Taylor and Francis Group.
9. Seeram E (2020) Computed Tomography Image Reconstruction. Radiologic Technology; (92), 2: 1–17.
10. Zicari RV (Ed) (2020). Explorations in Artificial Intelligence and Machine Learning. Boca Raton, FL, CRC Press: Taylor and Francis Group
11. Pesapane, F., Codari, M. & Sardanelli, F. (1918). Artificial intelligence in medical imaging: threat or opportunity? Radiologists again at the forefront of innovation in medicine. Eur Radiol Exp 2, 35. https://doi.org/10.1186/s41747-018-0061-6

12. Sheikh, H., Prins, C., Schrijvers, E. (2023). Artificial Intelligence: Definition and Background. In: Mission AI. Research for Policy. Springer, Cham. https://doi.org/10.1007/978-3-031-21448-6_2
13. IBM: What is Deep Learning? 2023. https://www.ibm.com/topics/deep-learning Accessed June 2023
14. Tobore I, Li J, Yuhang L, Al-Handarish Y, Kandwal A, Nie Z, Wang L (2019). Deep Learning Intervention for Health Care Challenges: Some Biomedical Domain Considerations. JMIR Mhealth Uhealth. 2019 Aug 2;7(8): e11966. https://doi.org/10.2196/11966. PMID: 31376272; PMCID: PMC6696854
15. Suzuki K. Overview of deep learning in medical imaging. Radiol Phys Technol. 2017;10(3):257–273. https://doi.org/10.1007/s12194-017-0406-5
16. Lee J-G, Jun S, Cho Y-W, Lee H, Kim GB, Seo JB, et al. Deep learning in medical imaging: general overview. Korean J Radiol; 2017; 18(4):570–84.
17. LeCun, Y.; Bengio, Y.; Hinton, G. Deep learning. (2015). Nature 521, 436–444.
18. Neural Networks. The Perceptron (2023). https://cs.stanford.edu/people/eroberts/courses/soco/projects/neuralnetworks/Neuron/index.html. Accessed June 2023
19. Jiang F, Jiang Y, Zhi H, Dong Y, Li H, Ma S, et al. (2017) Artificial intelligence in healthcare: past, present and future. Stroke Vasc Neurol; 2:230–43.
20. ESR-European Society of Radiology-Neri, E., de Souza, N. et al (2019). What the radiologist should know about artificial intelligence: an ESR white paper. Insights Imaging 10, 44 https://doi.org/10.1186/s13244-019-0738-2
21. Kim M, Yun J, Cho Y, Shin K, Jang R, Bae H-j, Kim N (2019) Deep Learning in Medical Imaging. Neurospine;16(4):657–668. https://doi.org/10.14245/ns.1938396.198
22. Rezazade Mehrizi MH, van Ooijen P, Homan, M. Applications of artificial intelligence (AI) in diagnostic radiology: a technography study. Eur Radiol 31, 1805–1811 (2021). https://doi.org/10.1007/s00330-020-07230-9
23. Kanade V (2023). What Is Artificial Intelligence (AI)? Definition, Types, Goals, Challenges, and Trends in 2022. https://www.spiceworks.com/tech/artificial-intelligence/articles/what-is-ai. Accessed July, 2023
24. Shvetsova N, Bakker B, Fedulova I, Schulz H, Dylov DV (2021). Anomaly Detection in Medical Imaging with Deep Perceptual Autoencoders," in IEEE Access, vol. 9, pp. 118571–118583, 2021, https://doi.org/10.1109/ACCESS.2021.3107163
25. Chalapathy R, Chawla S (2019) "Deep learning for anomaly detection: A survey", arXiv:1901.03407, 2019, [online] Available: http://arxiv.org/abs/1901.03407
26. Luo G, Xie W, Gao R, Zheng T, Chen L, Sun H (2023). Unsupervised anomaly detection in brain MRI: Learning abstract distribution from massive healthy brains, Computers in Biology and Medicine, Volume 154, 106610.
27. Valente J, António J, Mora C, Jardim S (2023). Developments in Image Processing Using Deep Learning and Reinforcement Learning. Journal of Imaging; 9(10):207. https://doi.org/10.3390/jimaging9100207
28. Hsieh, Jiang, Eugene Liu, Brian Nett, Jie Tang, Jean-Baptiste Thibault and Sonia Sahney. "A new era of image reconstruction: TrueFidelityTM Technical white paper on deep learning image reconstruction." (2019).
29. Boedeker K (2019). AiCE deep learning reconstruction: bringing the power of ultra-high-resolution CT to routine imaging. Canon Med Syst; 2:28–33.
30. What is Ethics? (2024). https://www.britannica.com/topic/ethics-philosophy Accessed Jan 2024
31. Brady AP, Neri E. Artificial Intelligence in Radiology (2020). Ethical Considerations. Diagnostics; 10(4):231. https://doi.org/10.3390/diagnostics10040231
32. Geis, J.R., Brady, A., Wu, C.C. et al. (2019). Ethics of artificial intelligence in radiology: summary of the joint European and North American multisociety statement. Insights Imaging 10, 101. https://doi.org/10.1186/s13244-019-0785-8

33. Artificial Intelligence in Medical Physics (2023). International Atomic Energy Agency (IAEA), Vienna International Centre, 1400 Vienna, Austria.
34. The Artificial Intelligence Era: The Role of Radiologic Technologists and Radiation Therapists (2020) ASRT Foundation, 15000 Central Ave. SE, Albuquerque, NM.
35. Artificial Intelligence and the Radiographer/Radiological Technologist Profession: A joint statement of the International Society of Radiographers and Radiological Technologists and the European Federation of Radiographer Societies. Guest Editorial (2020); Radiography 26: 93e-95
36. Rainey C, O'Regan T, Matthew J, Skelton E, Woznitza N, Chu K-Y, Goodman S, McConnell J, Hughes C, Bond R, McFadden S and Malamateniou C (2021) Beauty Is in the AI of the Beholder: Are We Ready for the Clinical Integration of Artificial Intelligence in Radiography? An Exploratory Analysis of Perceived AI Knowledge, Skills, Confidence, and Education Perspectives of UK Radiographers. Front. Digit. Health 3:739327, https://doi.org/10.3389/fdgth.2021.739327
37. Hardy M, Harvey H (2020) Artificial intelligence in diagnostic imaging: impact on the radiography profession. The British Journal of Radiology; 93(1108):20190840.
38. Malamateniou C. et al (2021) Artificial Intelligence: Guidance for clinical imaging and therapeutic radiography professionals, a summary by the Society of Radiographers AI working group. Radiography; 27: 1192–1202
39. Akudjedu TN, Torre S, Khine R, Katsifarakis D, Newm D, Malamateniou C (2023). Knowledge, perceptions, and expectations of Artificial intelligence in radiography practice: A global radiography workforce survey Journal of Medical Imaging and Radiation Sciences; 54:104–116
40. Yousef, R., Gupta, G., Yousef, N. et al. (2022). A holistic overview of deep learning approach in medical imaging. Multimedia Systems; 28, 881–914. https://doi.org/10.1007/s00530-021-00884-5

Chapter 2
AI Fundamentals

2.1 Introduction to Artificial Intelligence

Artificial intelligence (AI) is a multifaceted field that aims to replicate and augment human intelligence in machines, enabling them to reason and learn from experiences. Over the years, AI has evolved from being used solely in computer programs to being integrated into various products and services. For instance, AI-driven chatbots are revolutionizing customer service interactions, while autonomous vehicles are a testament to AI's role in transforming the automotive industry.

AI employs a diverse range of techniques, drawing from probability theory, economics, algorithm design, computer science, mathematics, psychology, and linguistics to solve practical problems. It leverages computer science tools for algorithm development and mathematical tools for modeling and solving optimization challenges.

The idea of AI dates back to the nineteenth century when Alan Turing introduced the concept of the "imitation game" as a means to assess machine intelligence [1]. Nonetheless, it has only become practically achievable in recent decades due to the significant advancements in computing capabilities and the increasing availability of vast datasets.

Human intelligence distinguishes itself by its capacity to learn from experiences and apply those lessons to novel situations. While today's computers don't match the complexity of the human brain, they excel at analyzing vast amounts of data rapidly. This ability enables AI to automate numerous complex tasks, from predicting maintenance needs to detecting fraud and optimizing delivery routes. This automation allows businesses to focus on core functions.

AI, as discussed in Chap. 1, is rooted in a diverse range of fields such as natural language processing (NLP), computer vision, neural computing, robotics, expert systems, and intelligent computer-assisted teaching, to name a few. It has seamlessly integrated into our daily lives, managing everything from emails and financial

E. Seeram, V. Kanade, *Artificial Intelligence in Medical Imaging Technology*,
https://doi.org/10.1007/978-3-031-64049-0_2

tasks to ecommerce, smart home systems, automobiles, IoT, and edge devices. Furthermore, AI is finding growing use in healthcare, medicine, and diverse medical imaging methods, including X-rays, CT scans, and MRI, among others.

2.2 Key Elements of AI

AI, with its foundations in a range of disciplines such as mathematics, cognitive science, and evolutionary theory, shows great potential for catalyzing future progress. Nonetheless, to be classified as AI, all its components must operate harmoniously. Key AI components have been illustrated in the diagram shown in Fig. 2.1.

Exploring these fundamental elements that make up AI will offer a thorough understanding:

1. **Machine learning (ML)**: ML is a subset of AI that focuses on training machines to learn from data. It involves algorithms that improve their performance over

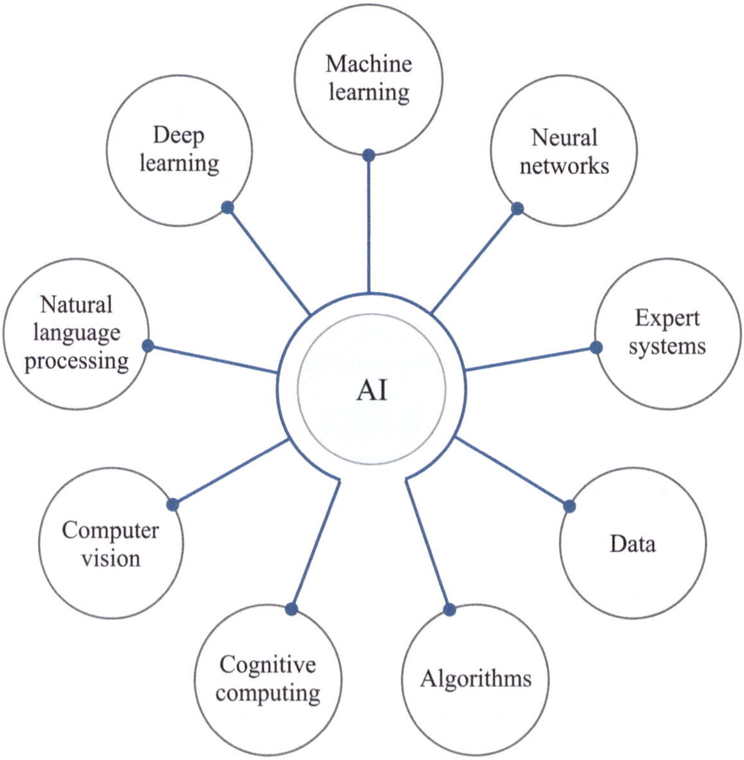

Fig. 2.1 Key elements of AI. See text for further explanation

time by recognizing patterns in data [2], for example, spam email filters that learn to distinguish between spam and legitimate emails.
2. **Deep learning (DL)**: DL is a subset of machine learning where artificial neural networks, inspired by the human brain, are used to process data [3]. It's particularly effective at tasks such as image recognition and natural language understanding due to its ability to automatically discover complex patterns in large datasets.
3. **Neural networks**: Neural networks are the building blocks of deep learning. They consist of interconnected nodes that mimic the structure of the human brain, enabling complex tasks such as facial or object recognition.
4. **Natural language processing**: NLP empowers machines to comprehend, decipher, and produce language spoken and written by humans. Examples include chatbots that converse with users and language translation apps such as Google Translate [4].
5. **Computer vision**: This component allows machines to interpret and understand visual information from images (i.e., graphs, tables, and PDF pictures) or videos. Self-driving cars use computer vision to perceive their surroundings and make driving decisions [5].
6. **Cognitive computing**: Cognitive computing strives to replicate the cognitive abilities of humans within a computer framework. Its goal is to simulate and enhance the interaction between humans and machines through the comprehension of human language and the interpretation of visual content [6].
7. **Expert systems**: These AI systems use knowledge bases and rule-based reasoning to solve specific problems or provide expert advice. Medical diagnostic systems and automated financial advisors are examples.
8. **Data**: Data is the fuel for AI. To achieve effective training, machine learning algorithms necessitate substantial datasets. Examples include the use of historical financial data to predict stock market trends and medical records for disease diagnosis.
9. **Algorithms**: AI algorithms are the mathematical instructions that guide AI systems' behavior and decision-making processes. For instance, reinforcement learning is a powerful algorithm that machines use to learn through trial and error by interacting with an environment and receiving rewards or penalties. An example is a self-driving car that might learn to navigate by getting positive feedback for safe driving and negative feedback for risky maneuvers, ultimately becoming proficient at driving autonomously.

AI's diverse components come together in various applications across industries, from autonomous vehicles and virtual personal assistants to fraud detection systems and recommendation engines, revolutionizing how businesses operate and enhancing our daily lives.

2.3 Definition of AI and Its Applications in Healthcare

This section will cover the applications of AI in healthcare.

2.3.1 *Artificial Intelligence*

Artificial intelligence (AI) refers to the field of computer science and technology that aims to create systems and machines capable of performing tasks that typically require human intelligence, such as learning, reasoning, problem-solving, and adapting to new information and environments. At its fundamental essence, AI systems are engineered with the primary purpose of ingesting and scrutinizing data, diligently discerning intricate patterns within it, and subsequently leveraging this acquired knowledge to formulate predictions or recommendations.

These AI-driven mechanisms, characterized by their adaptability and capacity for learning, serve to automate tasks, solve complex problems, and emulate human-like cognitive functions. They contribute to a myriad of applications across industries and significantly enhance our ability to process information, optimize decision-making, and redefine various facets of our daily lives.

2.3.2 *Applications of AI in Healthcare*

Artificial intelligence has reshaped healthcare by offering innovative solutions to long-standing challenges. Its applications in disease diagnosis, medical imaging, drug discovery, personalized medicine, virtual health assistants, and administrative tasks have the potential to improve patient outcomes, reduce costs, and enhance the overall quality of healthcare. To comprehend some of the key applications of AI in healthcare, one must delve into various aspects of its implementation.

1. **Disease Diagnosis and Risk Prediction**

   AI's ability to analyze vast amounts of medical data has significantly improved disease diagnosis and risk prediction. For instance, Google's DeepMind developed an AI system that can analyze retinal scans to detect diabetic retinopathy, a leading cause of blindness [7]. The system's accuracy in diagnosing the condition matches that of human experts, demonstrating its potential in early disease detection.

   Furthermore, AI-powered predictive models can assess a patient's risk of developing various diseases, such as heart disease, cancer, and diabetes. These models consider a wide range of factors, including genetic data, lifestyle choices, and medical history, to provide personalized risk assessments and preventive recommendations.

2. **Medical Imaging and Radiology**
   AI has revolutionized medical imaging by enhancing the accuracy and speed of image analysis. For example, IBM's Watson for Oncology employs AI to analyze medical images, pathology reports, and medical literature to assist oncologists in creating personalized treatment plans for cancer patients [8]. AI algorithms can quickly detect anomalies in X-rays, MRIs, and CT scans, aiding in early detection of conditions such as tumors, fractures, and infections.
3. **Drug Discovery and Development**
   The pharmaceutical industry has embraced AI to expedite drug discovery and development processes. AI algorithms possess the capability to scrutinize extensive repositories of molecular configurations, genetic data, and clinical trial records, with the aim of pinpointing potential pharmaceutical candidates. Atomwise, a company specializing in AI-driven drug discovery, uses deep learning models to predict how molecules will interact with disease targets, potentially accelerating the development of new medications [9].
4. **Personalized Medicine**
   Personalized medicine aims to provide tailored treatment plans based on an individual's unique genetic makeup and medical history. AI plays a pivotal role in this endeavor by analyzing genetic data to identify optimal treatment options and predict potential adverse reactions. One notable example is the use of AI in cancer treatment, where algorithms analyze tumor genetics to recommend personalized therapies that target specific mutations.
5. **Virtual Health Assistants**
   AI-driven virtual health assistants and chatbots such as Babylon Health and Ada Health have gained popularity for their ability to provide healthcare information, schedule appointments, and monitor patients' health remotely [45, 46]. These applications have been particularly beneficial during the COVID-19 pandemic, as they reduce the burden on healthcare professionals and enable patients to access care from the safety of their homes [10].
6. **Administrative Efficiency**
   AI is also being used to streamline administrative tasks in healthcare, such as billing, claims processing, and appointment scheduling. NLP algorithms can analyze medical documents and extract relevant information, reducing administrative workload and minimizing errors [11]. This enhanced efficiency empowers healthcare professionals to dedicate a greater portion of their attention and efforts to patient well-being.

   Researchers at top universities, such as Stanford University, are already developing AI algorithms that can predict patient deterioration in real time by analyzing electronic health records [12]. Such a system can help medical staff intervene promptly when a patient's condition worsens, potentially saving lives.
7. **Robotic Surgery**
   AI plays a crucial role in robotic surgery by enhancing the precision, efficiency, and safety of surgical procedures. In robotic surgery, AI can assist in several ways. Firstly, it can help surgeons with preoperative planning by analyzing

patient data and generating personalized surgical plans. During surgery, AI-powered systems can provide real-time feedback to surgeons, such as identifying critical structures and suggesting optimal incision points. AI can also enhance the capabilities of robotic surgical instruments, making them more dexterous and adaptive. Additionally, AI can assist in postoperative monitoring and analysis to ensure patients' recovery progresses smoothly.

A recent example of AI in robotic surgery is the da Vinci Surgical System, which has been widely adopted in various surgical specialties [13]. This robotic platform incorporates AI to improve surgical precision by filtering out tremors and providing 3D visualization. Moreover, the system can assist surgeons by translating their hand movements into more precise robotic actions, reducing the risk of errors and complications during procedures [14]. Such advancements in robotic surgery, driven by AI, continue to disrupt the field and offer patients safer and more effective treatment options.

With ongoing research and development, AI will continue to play a pivotal role in reforming the healthcare landscape, ultimately leading to more efficient and effective healthcare delivery.

2.4 History and Evolution of AI

The history of AI is a fascinating journey that spans over several decades, with its roots tracing back to the mid-twentieth century. The concept of AI was first introduced by philosophers who attempted to describe human thinking as the mechanical manipulation of symbols. However, the field of AI research was officially founded at a workshop held at Dartmouth College in 1956, where the term "artificial intelligence" was coined [15].

The journey of AI started in the 1950s with the pioneering work of Alan Turing, who proposed the Turing test to determine if a machine could mimic human intelligence [1]. In the same decade, John McCarthy made pioneering strides in the field of AI by conceiving Language Instructional System Planning (LISP), the first-ever AI programming language, during his tenure at MIT School. This language represented a groundbreaking achievement in the realm of AI research [16].

The 1960s and 1970s marked the first wave of AI development, with the creation of programs such as the General Problem Solver (GPS) and expert systems [17]. These systems could solve specific problems or offer expert-level advice within limited domains. However, they struggled when faced with real-world complexity due to their rigid rule-based nature. Early AI systems focused on symbolic reasoning and rule-based systems, leading to the development of expert systems in the 1970s and 1980s.

The introduction of principles of probability in AI research in 1988 marked a significant milestone, as it allowed for more sophisticated AI models to be developed. The 1990s and early 2000s saw significant advancements in AI, such as IBM's Deep Blue defeating the world chess champion Garry Kasparov in 1997 [18].

The twenty-first century witnessed a boom in AI research and applications, driven by the successful application of machine learning to many problems in academia and industry, thanks to new methods, powerful computer hardware, and the collection of immense datasets. The launch of the ImageNet Large Scale Visual Recognition Challenge (ILSVRC) in 2010 marked another important milestone, spurring further advancements in AI technologies [19].

In recent years, the evolution of AI has been nothing short of astonishing. Machine learning, a subset of AI, has given rise to remarkable achievements. For instance, in the realm of computer vision, AI-driven systems can now generate photorealistic images that are nearly indistinguishable from those captured by a camera. These advancements have found applications in numerous creative fields, from art and design to video game development and even filmmaking.

AI has indeed made remarkable strides, surpassing human capabilities in numerous fields. For instance, in 2015, AlphaGo achieved the historic feat of becoming the first AI to defeat a professional Go player [20]. Furthermore, self-driving cars equipped with advanced AI systems, such as those developed by Tesla and Waymo, have shown the potential to navigate complex road environments more safely and efficiently than human drivers [21].

AI has also shaken up healthcare, where it's being used to analyze medical images, such as X-rays and MRIs, with remarkable precision. This not only speeds up diagnoses but can also detect diseases at earlier stages when they're more treatable. Moreover, AI has become integral in industries such as manufacturing, where it's optimizing production processes and quality control, thereby increasing efficiency and reducing errors.

Furthermore, NLP has reached a level where AI systems can interpret and generate human language with remarkable fluency. Virtual assistants such as Siri and Alexa have become household staples, answering queries and performing tasks based on voice commands [22]. Beyond that, language models such as GPT-3 can generate humanlike text, from creative writing and content generation to language translation and chatbots [23]. In education, AI-driven platforms are personalizing learning experiences, adapting content to individual students' needs, and providing instant feedback.

These developments illustrate how AI has evolved over time and woven itself into the fabric of our daily lives, reshaping communication, education, medicine, and the way we interact with information and technology. The relentless pursuit of knowledge and innovation in AI continues to redefine the possibilities of what machines can achieve, promising a future filled with even more remarkable applications and transformative impact.

2.5 Overview of Machine Learning and Deep Learning

Machine learning (ML) and deep learning (DL) are two powerful subsets of AI that are breaking new grounds across various industries, including healthcare. These technologies are transforming the way professionals in different fields approach problem-solving, decision-making, and service delivery.

2.5.1 Machine Learning Overview

Machine learning represents a facet of artificial intelligence that harnesses statistical techniques to empower computers with the capacity to learn and make decisions autonomously, without explicit programming. This paradigm is rooted in the concept that computers possess the ability to assimilate knowledge from data, detect patterns, and exercise judgment with minimal human intervention.

Within the broader domain of artificial intelligence, machine learning endeavors to imbue machines with humanlike behaviors and decision-making capabilities by endowing them with the capacity to self-learn and evolve their own operational algorithms. This process unfolds with a notable absence of explicit programming, as automation takes center stage and the machine's performance advances based on its accumulated experiences.

In practice, machine learning begins with the provisioning of high-quality data to the machine, followed by the application of various algorithms to construct machine learning models for training. The selection of an algorithm hinges upon the nature of the data at hand and the specific task targeted for automation.

It's worth distinguishing machine learning from traditional programming: In traditional programming, we typically supply input data alongside a meticulously crafted and rigorously tested program to produce desired outputs. In contrast, machine learning involves a learning phase where both input data and corresponding output are presented to the machine, allowing it to formulate its own program. To illustrate this distinction more vividly, please refer to the accompanying Fig. 2.2.

Although the concept of machine learning traces back to World War II, when the Enigma machine was employed, the automation of intricate mathematical computations on increasingly vast and diverse datasets is a relatively recent advancement.

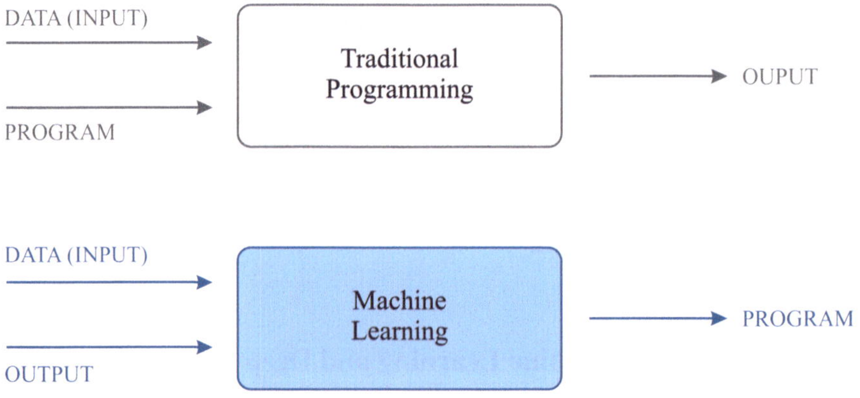

Fig. 2.2 Traditional versus ML programming. See text for further explanation

In today's era characterized by the proliferation of big data, IoT, and pervasive computing, machine learning has become indispensable for addressing a wide array of challenges in fields such as:

1. Computational finance (including credit scoring and algorithmic trading)
2. Computer vision (encompassing facial recognition, motion tracking, and object detection)
3. Computational biology (spanning DNA sequencing, brain tumor identification, and drug discovery)
4. Automotive, aerospace, and manufacturing (notably in predictive maintenance)
5. Natural language processing (with applications like voice recognition)

2.5.2 How ML Works?

Machine learning involves instructing a computer to learn from past experiences, much like individuals draw from their own history to make decisions. Consider the scenario of a radiologist examining MRI images. Through exposure to numerous MRI scans, the radiologist gains an understanding of what constitutes a healthy or unhealthy image. Subsequently, when presented with a new MRI scan, past knowledge is applied to formulate a diagnosis. Machine learning parallels this process but on a broader scale, utilizing extensive datasets and mathematical algorithms.

Think of the learning curve in a different context: Initially interpreting MRI scans is comparable to attempting to play the piano without prior lessons. Without a clear understanding of crucial aspects, interpretations may be random. However, with exposure to more scans and learning from mistakes, proficiency in identifying abnormalities improves.

In the realm of machine learning, the computer is trained to analyze MRI scans. While its initial interpretations may be arbitrary, continuous exposure to and learning from a multitude of MRI scans enhance its ability to make precise diagnoses. This is akin to honing piano-playing skills through consistent practice. Now, consider another medical imaging example: Picture having a database of medical images, such as X-rays, alongside corresponding diagnoses:

- Image A—Shows a healthy lung, labeled as "Normal".
- Image B—Displays an X-ray of a lung with pneumonia, labeled as "Pneumonia".
- Image C—Represents a lung affected by lung cancer, labeled as "Cancer".

When presenting a machine learning model with images and their corresponding diagnoses, the model can learn to identify patterns within the images. Machine learning works with more intricate data, specifically the detailed information within medical images in this case.

Machine learning models analyze numerous medical images and their diagnoses. Sophisticated algorithms are employed to identify subtle patterns, such as the presence of abnormalities or diseases. When inputting a new medical image, the machine learning model utilizes the patterns it has learned to make an informed prediction about the presence or absence of a medical condition, assisting healthcare professionals in making accurate diagnoses and treatment decisions.

In short, machine learning is like teaching a computer to learn from examples and use that knowledge to make predictions. It's kind of like teaching it to play a musical instrument or analyzing temperature and ice cream sales data to make forecasts, but it uses math to do it. In medicine, it helps doctors and radiologists make better decisions by analyzing lots of patient data and finding hidden patterns.

2.5.3 Deep Learning Overview

Deep learning, a subset of artificial intelligence and machine learning, mimics human learning processes to extract knowledge from various data types. These deep learning models are adept at tasks such as recognizing patterns in images, processing text, and analyzing audio data, as well as automating tasks typically requiring human intelligence.

In the realm of medical imaging, deep learning models, including artificial neural networks (ANNs) and convolutional neural networks (CNNs), have revolutionized medical imaging [24]. These models are designed to process pixel data within images, leading to enhanced diagnostic precision. Among their applications, CNNs have been instrumental in reconstructing CT images, more crucially in low-dose CT (LDCT) scans where minimizing radiation exposure is of high priority [25].

In the quest to reduce image noise and improve the quality of calculated tomography scans, CNN-based algorithms known as deep learning image reconstruction (DLIR) algorithms have been utilized [26]. These are especially valuable in LDCT lung cancer screening, where clearer images are obtained with lower radiation doses, thus improving patient safety and diagnosis.

Neural networks and ANNs resemble neurons in the human brain in functioning. They rely on artificial neurons trained via algorithms inspired by the learning process of the human brain. To perform complex tasks such as medical image analysis and reconstruction, the AI system uses the training process in which input signals and desired output decisions are presented to the network, thereby adjusting internal parameters or weights to map inputs to desired outputs.

This iterative mechanism is what makes it possible for ANNs, including CNNs, to handle tasks such as image denoising, segmentation, and disease detection. The fusion of deep learning tech and medical imaging could reshape diagnostic practices, leading to earlier disease discovery, more accurate diagnoses, and improved patient outcomes, showing an increasingly promising future in global healthcare.

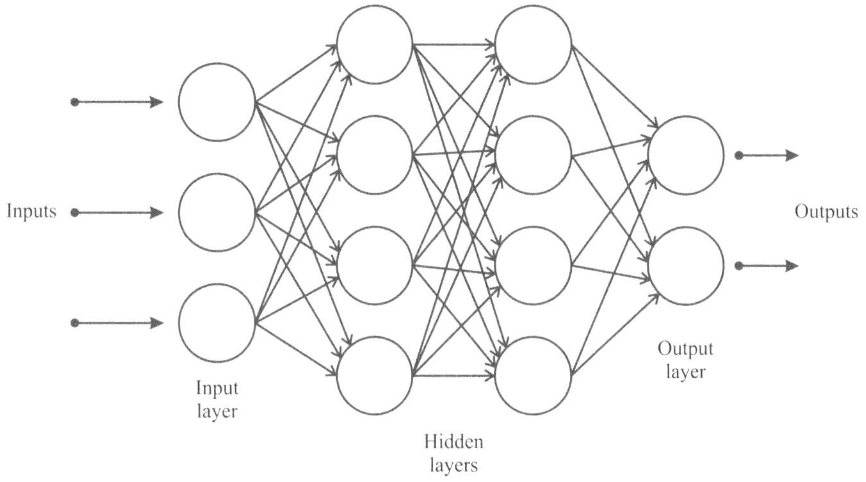

Fig. 2.3 Multilayered neural networks: Interconnected nodes in structured layers. See text for further explanation

2.5.4 How Deep Learning Works?

To grasp the concept of deep learning, envision a young child learning the word "dog." The child associates objects with the word "dog," and a parent confirms whether the object is indeed a dog or not. Over time, the child accumulates knowledge about the common characteristics of dogs. Unbeknownst to the child, they are forming a hierarchical understanding, where each layer of abstraction builds upon the knowledge gained from the previous one.

Deep learning algorithms, such as those used in computer programs, undergo a process akin to a child learning to recognize a dog. Imagine a deep learning program as a multifaceted structure with interconnected layers, each layer building upon the previous one to enhance predictions and classifications. These layers execute intricate, nonlinear transformations on input data and employ their newfound knowledge to generate a statistical model as the output. This iterative process continues until the output reaches an acceptable level of precision. The "deep" in deep learning denotes the numerous layers data must traverse. Refer to Fig. 2.3 for better understanding.

Thus, deep learning is a subfield of machine learning that focuses on neural networks with multiple layers, known as deep neural networks. These networks are designed to automatically learn and extract complex features from data through a hierarchy of layers.

Illustrating the concept of deep learning involves considering the task of teaching a computer to recognize handwritten numbers, such as those commonly found on medical forms or X-ray labels.

Consider a collection of handwritten numbers from 0 to 9 in a large pile. Initially, they may appear as messy scribbles, similar to an initial perception. Deep learning operates somewhat like the process of distinguishing these numbers.

In deep learning, the computer has many layers, such as layers of detectives. The first detectives look at simple things, such as straight lines and curves. Then, the next detectives look at more complicated stuff, such as loops or slanted lines. Finally, the last detectives put all their clues together and can say which number it is.

When a computer is presented with a messy handwritten number, it initially identifies the lines and curves, progressing to more intricate features such as loops and slants. Finally, it says, "Oh, this is the number 5!" Deep learning helps computers do this for many different numbers and letters.

Now, consider medical images. Imagine a scenario where a computer needs to identify various body parts in X-rays, such as bones or organs.

At first, the computer doesn't know what it's looking at. But, just like with the numbers, it starts with simple things, like edges and shapes. Then, as it looks deeper, it recognizes more complex things, such as the shape of a specific bone or organ. Ultimately, the system can convey what it observes in the X-ray.

Deep learning is like training the computer to become a super detective for medical images. It looks at all the details, from simple to complex, to help radiologists and imaging technologists find and understand what's in the X-ray.

So, deep learning is a way to teach computers to see and understand things in pictures and text, just like we learn to recognize numbers or body parts by looking at different details. It's really useful for tasks such as spotting diseases in medical images.

2.6 Clinical Decision Support Systems

Artificial intelligence is reinventing clinical decision support systems (CDSSs) in the field of healthcare in numerous profound ways [27]. CDSS, traditionally reliant on manual analysis of patient data and medical literature, now benefits from AI's capacity to swiftly process vast volumes of information and identify patterns, ultimately enhancing diagnostic accuracy and treatment recommendations. Machine learning algorithms are adept at predicting disease outcomes, assisting in risk assessment, and suggesting tailored treatment plans by analyzing patient data, such as medical records, imaging results, and genomic information.

Furthermore, AI-driven CDSS can offer real-time alerts and reminders to clinicians, improving adherence to best practices and reducing errors. By continuously learning and adapting from new data, AI-powered CDSS evolves to incorporate the latest medical knowledge and research findings, ensuring that healthcare providers have access to the most up-to-date information for informed decision-making.

Despite these remarkable advancements, it is crucial to maintain a balance between AI and human expertise, as AI should be viewed as a valuable tool that complements and enhances the clinical decision-making process rather than replacing human healthcare professionals.

2.6.1 Integration of AI Algorithms into Clinical Workflows

Recent advancements in AI technologies have transformed medical imaging by integrating AI algorithms into clinical workflows. This integration provides radiologists and medical professionals with advanced tools capable of rapidly and accurately processing extensive volumes of medical images.

For instance, companies such as Aidoc and Zebra Medical Vision have developed AI-driven platforms that seamlessly integrate with radiology systems [28, 29]. These platforms automatically detect anomalies, such as fractures or tumors, within medical images. Radiologists can efficiently prioritize cases, ensuring that critical ones receive immediate attention. These AI-assisted tools enhance the capabilities of radiologists, improve diagnostic precision, and reduce the likelihood of oversight.

Furthermore, the integration of AI algorithms not only benefits patient care but also contributes to the accumulation of invaluable data, which can be harnessed for research and the development of more effective diagnostic and treatment approaches. This wealth of data can potentially lead to groundbreaking discoveries in medical science and further advancements in patient care, marking a profound transformation in the healthcare landscape.

2.6.2 Challenges and Ethical Considerations in Implementing AI Systems in Healthcare

Implementing AI systems in healthcare presents several challenges and ethical dilemmas. One of the most prominent concerns revolves around data privacy and security. AI algorithms require extensive datasets for training, often containing sensitive patient information. It is imperative that healthcare organizations adhere strictly to data protection regulations, such as HIPAA (in the United States) and GDPR (in the European Union), to safeguard patient data [30, 31].

Another ethical consideration pertains to the transparency of AI algorithms. Radiologists and patients must comprehend how AI systems reach their conclusions. Deep learning models are frequently described as "black boxes" due to their complexity, rendering it difficult to trace the decision-making process. Ensuring transparency and interpretability is essential to establish trust in AI-supported diagnoses.

Furthermore, AI systems may inadvertently perpetuate biases present in their training data, leading to disparities in healthcare outcomes. For example, a study conducted by Obermeyer et al. (2019) unveiled racial bias in a widely used AI algorithm for healthcare, favoring white patients over black patients [32]. Radiologists and medical professionals must actively address and mitigate such biases to ensure equitable patient care.

2.6.3 Interpreting and Validating AI Results for Clinical Decision-Making

Interpreting and validating AI results for clinical decision-making is a critical aspect of integrating artificial intelligence into healthcare. While AI has shown great promise in improving diagnostic accuracy and treatment recommendations, it's imperative to ensure that the results are reliable and understandable for healthcare professionals.

One essential aspect of this process is interpretability, which involves making AI models and their predictions transparent and comprehensible. Recent advancements in AI research have focused on developing interpretable models that provide clinicians with insights into why a particular decision or recommendation was made. For example, techniques such as attention mechanisms in deep learning models can highlight the most relevant features in medical images or patient records, helping physicians understand the rationale behind a diagnosis or treatment suggestion [33]. This transparency builds trust and confidence in AI-driven clinical decision support systems.

Moreover, validating AI results is crucial to assess their accuracy and reliability. Recent examples include studies that validate AI algorithms for various medical tasks, such as detecting diabetic retinopathy in retinal images or predicting patient outcomes based on electronic health records. These validation studies often involve large datasets and rigorous evaluation methods to ensure that AI models perform consistently and meet the required standards for clinical use.

Furthermore, regulatory bodies such as the US Food and Drug Administration (FDA) have started providing guidelines and pathways for the approval of AI-based medical devices, emphasizing the importance of rigorous validation before these technologies are deployed in clinical settings [34].

In summary, ensuring the interpretability and validation of AI results in clinical decision-making is vital to harness the full potential of AI while maintaining patient safety and healthcare quality. Recent developments in AI research and regulatory practices are helping pave the way for responsible and effective integration of AI in healthcare.

2.7 Importance of AI in Medical Imaging

AI in medical imaging is a rapidly evolving field with tremendous potential to improve healthcare outcomes by enabling earlier and more accurate diagnoses, streamlining workflows, personalizing treatment plans, and advancing medical research.

AI-driven technologies in medical imaging are helping healthcare providers make more accurate and timely decisions, leading to better patient outcomes. Here's a comprehensive explanation with recent examples apart from the mentioned companies:

1. **Early detection and diagnosis**: AI can analyze medical images with incredible speed and precision, allowing for the early detection of diseases and conditions. For example, in breast cancer screening, AI algorithms can identify abnormalities in mammograms that may be missed by human radiologists. Companies such as NVIDIA have developed AI tools for this purpose. Their Clara Breast AI is designed to assist radiologists in breast cancer screening [35].
2. **Efficient workflow**: AI can streamline the workflow of healthcare professionals. For instance, Arterys, a medical imaging AI company, has developed a platform that uses deep learning to analyze cardiac MRI scans [36]. It reduces the time it takes to perform the analysis from hours to just a few minutes, allowing doctors to make quicker decisions.
3. **Personalized treatment plans**: AI can help create personalized treatment plans based on a patient's medical images and health data. PathAI, for example, uses AI to assist pathologists in diagnosing diseases from pathology slides [37]. It can provide insights into which treatment options are likely to be most effective for a particular patient.
4. **Radiomics and predictive analytics**: AI's ability to extract quantitative data from medical images (radiomics) and integrate it with clinical information empowers predictive analytics. Predictive analytics can forecast disease progression, treatment outcomes, and patient response to therapies. For example, Paige.AI leverages AI to build predictive models for cancer patients, helping oncologists make informed decisions about treatment strategies [38].
5. **Reducing errors**: AI can reduce the chances of human error in medical imaging. For instance, 'Infervision' has developed an AI system for radiologists that can detect and highlight abnormalities in chest X-rays [39]. This can help radiologists catch potential issues they might have missed.
6. **Telemedicine**: AI can play a vital role in telemedicine, especially in remote areas. Telemedicine platforms such as Arterys use AI to provide remote access to advanced imaging analysis, enabling healthcare providers in underserved areas to access expert opinions and improve patient care. Moreover, companies such as Caption Health have created AI solutions that guide healthcare providers in capturing high-quality ultrasound images, even if they are not ultrasound experts [40]. This helps in expanding access to medical imaging expertise.
7. **Research and drug development**: AI can analyze vast amounts of medical imaging data to aid in research and drug development. PathAI, for instance, assists not only in diagnosis but also in research efforts to understand diseases at a deeper level.
8. **Resource optimization**: AI can help optimize the allocation of resources in healthcare facilities. For example, MaxQ AI has developed AI algorithms that can quickly analyze CT scans of the brain for signs of stroke [41]. This helps healthcare providers prioritize urgent cases and allocate resources accordingly.

9. **Patient engagement**: AI can enhance patient engagement by providing visualizations and explanations of medical images to patients. Companies such as HeartFlow use AI to create 3D models of coronary arteries, helping patients better understand their heart conditions and treatment options [42].
10. **Triage and resource allocation**: During public health emergencies or natural disasters, AI can play a crucial role in triaging patients. For instance, AI systems can analyze chest X-rays to prioritize COVID-19 cases, ensuring that resources are allocated efficiently during surges in patient volume.
11. **Global access to expertise**: AI democratizes access to medical expertise and diagnostic services. EchoNous, for instance, developed an AI-driven ultrasound device that can be operated by healthcare professionals with minimal training [43]. This portable technology enables ultrasound imaging in remote or underserved areas, improving healthcare accessibility and early disease detection where resources are limited. Thus, AI can bridge the gap in access to specialized medical expertise.
12. **Education and training**: AI aids in medical education by offering valuable teaching tools. For example, the 3D4Medical platform uses AI to create detailed 3D models of anatomical structures, allowing medical students and professionals to explore the human body in unprecedented detail [44]. These interactive models enhance learning and help individuals grasp complex concepts, ultimately contributing to better-prepared healthcare providers.

Thus, the importance of AI in medical imaging cannot be overstated. It enhances diagnostic accuracy, streamlines workflows, enables early intervention, personalizes treatment, reduces costs, and expands access to expertise. These advancements are not hypothetical; they are already making a significant impact in various medical fields, ultimately leading to better patient outcomes and improved healthcare globally. As AI continues to evolve, its role in medical imaging will only become more prominent, ushering in a new era of healthcare.

References

1. Oppy, Graham and David Dowe, "The Turing Test", The Stanford Encyclopedia of Philosophy (Winter 2021 Edition), Edward N. Zalta (ed.), URL = https://plato.stanford.edu/archives/win2021/entries/turing-test/
2. Kanade V., What Is Machine Learning? Definition, Types, Applications, and Trends for 2022. URL: https://www.spiceworks.com/tech/artificial-intelligence/articles/what-is-ml/#:~:text=Machine%20learning%20teaches%20machines%20to,Vijay%20Kanade%20AI%20Researcher
3. IBM, What is deep learning? URL: https://www.ibm.com/topics/deep-learning
4. Google Research, Natural Language Processing, URL: https://research.google/research-areas/natural-language-processing/
5. B. Asmika, G. Mounika and P. S. Rani, "Deep Learning for Vision and Decision Making in Self Driving Cars-Challenges with Ethical Decision Making," 2021 International Conference on Intelligent Technologies (CONIT), Hubli, India, 2021, pp. 1–5, https://doi.org/10.1109/CONIT51480.2021.9498342.

6. M. Chen, F. Herrera and K. Hwang, "Cognitive Computing: Architecture, Technologies and Intelligent Applications," in IEEE Access, vol. 6, pp. 19774–19783, 2018, https://doi.org/10.1109/ACCESS.2018.2791469.
7. Cade Metz, Google's AI Reads Retinas to Prevent Blindness in Diabetics, Nov. 29, 2016, URL: https://www.wired.com/2016/11/googles-ai-reads-retinas-prevent-blindness-diabetics/
8. Liu C, Liu X, Wu F, Xie M, Feng Y, Hu C. Using Artificial Intelligence (Watson for Oncology) for Treatment Recommendations Amongst Chinese Patients with Lung Cancer: Feasibility Study. J Med Internet Res. 2018 Sep 25;20(9):e11087. https://doi.org/10.2196/11087. PMID: 30257820; PMCID: PMC6231834.
9. Atomwise, We Are Reinventing Small Molecule Drug Discovery, URL: https://www.atomwise.com/how-we-do-it/
10. Sarah Moore, Beyond Virtual Consultations: The Rise of AI-Driven Virtual Healthcare Assistants, Jan 30, 2024, URL: https://www.news-medical.net/health/Beyond-Virtual-Consultations-The-Rise-of-AI-Driven-Virtual-Healthcare-Assistants.aspx
11. Natural Language Processing in Healthcare: Using Text Analysis for Medical Documentation and Decision-Making, October 25, 2021, URL: https://www.altexsoft.com/blog/nlp-healthcare/
12. Healthcare, Life Sciences, and AI, March 2021,URL: https://hai.stanford.edu/sites/default/files/2022-01/HAI_IndustryBrief1_%5B56%5D.pdf
13. The Da Vinci Technology: Pioneering a New Era in Medical Imaging and Patient Care, URL: https://openmedscience.com/the-da-vinci-technology-pioneering-a-new-era-in-medical-imaging-and-patient-care/#:~:text=The%20incorporation%20of%20AI%20within,patient%20experiences%20and%20recovery%20times.
14. Kanade, V.A. Light Coded Language (LCL): Infrared-Thermography* based 3D Interactive Touch Sensing Display for Handling Medical Emergencies. SN COMPUT. SCI. 1, 147 (2020). https://doi.org/10.1007/s42979-020-00172-5
15. Artificial Intelligence Coined at Dartmouth, URL: https://home.dartmouth.edu/about/artificial-intelligence-ai-coined-dartmouth
16. The History of Artificial Intelligence, History of Computing, CSEP 590A, University of Washington, December 2006, URL: https://courses.cs.washington.edu/courses/csep590/06au/projects/history-ai.pdf
17. Edem Gold, The History of Artificial Intelligence from the 1950s to Today, April 13, 2023, https://www.freecodecamp.org/news/the-history-of-ai/
18. IBM, Deep Blue, URL: https://www.ibm.com/history/deep-blue
19. Jason Brownlee, A Gentle Introduction to the ImageNet Challenge (ILSVRC), July 5, 2019, URL: https://machinelearningmastery.com/introduction-to-the-imagenet-large-scale-visual-recognition-challenge-ilsvrc/
20. Google DeepMind, AlphaGo, https://deepmind.google/technologies/alphago/
21. Sean O'Kane, How Tesla and Waymo are tackling a major problem for self-driving cars: data, April 19, 2018, URL: https://www.theverge.com/transportation/2018/4/19/17204044/tesla-waymo-self-driving-car-data-simulation
22. ZDNET, The best voice assistant, September 09, 2021, URL: https://www.zdnet.com/home-and-office/smart-home/the-best-voice-assistant/
23. Language Models, Explained: How GPT and Other Models Work, January 18, 2023, URL: https://www.altexsoft.com/blog/language-models-gpt/
24. Kim M, Yun J, Cho Y, Shin K, Jang R, Bae HJ, Kim N. Deep Learning in Medical Imaging. Neurospine. 2019 Dec;16(4):657–668. https://doi.org/10.14245/ns.1938396.198. Epub 2019 Dec 31. Erratum in: Neurospine. 2020 Jun;17(2):471–472. PMID: 31905454; PMCID: PMC6945006.
25. Danna, S. Jérôme. (October 2020). Deep Learning Enables CT Image Reconstruction at Low Dosages. Radiological Society of North America. https://www.rsna.org/news/2020/October/AI-CT-Image-Reconstruction
26. Zhong J, Xia Y, Chen Y, Li J, Lu W, Shi X, Feng J, Yan F, Yao W, Zhang H. Deep learning image reconstruction algorithm reduces image noise while alters radiomics features in dual-energy CT in comparison with conventional iterative reconstruction algorithms: a phantom study. Eur Radiol. 2023 Feb;33(2):812–824. https://doi.org/10.1007/s00330-022-09119-1. Epub 2022 Oct 5.

27. Wang L, Chen X, Zhang L, Li L, Huang Y, Sun Y, Yuan X. Artificial intelligence in clinical decision support systems for oncology. Int J Med Sci. 2023 Jan 1;20(1):79–86. https://doi.org/10.7150/ijms.77205. PMID: 36619220; PMCID: PMC9812798.
28. Radiologists' Go-to AI Solution, URL: https://www.aidoc.com/solutions/radiology/
29. Zebra Medical Vision AI1 (All-in-One) solution transforms patient care with the power of AI, January 01, 2020, https://whatsnext.nuance.com/healthcare-ai/zebra-medical-vision-ai1-transforms-patient-care/
30. CDC, Health Insurance Portability and Accountability Act of 1996 (HIPAA), URL: https://www.cdc.gov/phlp/publications/topic/hipaa.html#:~:text=Health%20Insurance%20Portability%20and%20Accountability%20Act%20of%201996%20(HIPAA),-Print&text=The%20Health%20Insurance%20Portability%20and,the%20patient's%20consent%20or%20knowledge.
31. European Council, The general data protection regulation, URL: https://www.consilium.europa.eu/en/policies/data-protection/data-protection-regulation/#:~:text=data%20protection%20rules-,What%20is%20the%20GDPR%3F,application%20on%2025%20May%202018.
32. Obermeyer Z, Powers B, Vogeli C, Mullainathan S. Dissecting racial bias in an algorithm used to manage the health of populations. Science. 2019 Oct 25;366(6464):447–453. https://doi.org/10.1126/science.aax2342.
33. Li, X., Li, M., Yan, P., Li, G., Jiang, Y., Luo, H., & Yin, S. Deep Learning Attention Mechanism in Medical Image Analysis: Basics and Beyonds. International Journal of Network Dynamics and Intelligence. 2023, 2(1), 93–116. https://doi.org/10.53941/ijndi0201006.
34. U.S. FDA, Artificial Intelligence and Machine Learning in Software as a Medical Device, URL: https://www.fda.gov/medical-devices/software-medical-device-samd/artificial-intelligence-and-machine-learning-software-medical-device
35. AI Model Matches Radiologists' Accuracy Identifying Breast Cancer in MRIs, September 28, 2022, URL: https://developer.nvidia.com/blog/ai-model-matches-radiologists-accuracy-identifying-breast-cancer-in-mris/
36. Startup Brings AI-Powered Image Analysis to Heart, Liver and Lungs, November 7, 2018, URL: https://blogs.nvidia.com/blog/arterys-ai-powered-image-analysis/
37. Improving Patient Outcomes with AI-Powered Pathology, URL: https://www.pathai.com/
38. Bringing the power of generative AI across the cancer continuum, URL: https://paige.ai/
39. Erik L. Ridley, AI start-up eyes deep learning to aid imaging in China, May 7, 2017, URL: https://www.auntminnie.com/imaging-informatics/artificial-intelligence/article/15617528/ai-start-up-eyes-deep-learning-to-aid-imaging-in-china
40. Solving early disease detection with AI, URL: https://www.caption-care.com/technology#:~:text=Now%2C%20with%20Caption%20AI%2C%20any,key%20measure%20of%20heart%20health.
41. Conor Hale, FDA clears MaxQ's AI-based CT analysis platform for brain bleeds, Nov 9, 2018, URL: https://www.fiercebiotech.com/medtech/fda-clears-maxq-s-ai-based-ct-analysis-platform-for-brain-bleeds
42. HeartFlow Analysis is First AI-Enabled Technology to be Recognized by the American College of Cardiology and American Heart Association Guidelines as Important Tool in Diagnosing and Treating Heart Disease, October 28, 2021, URL: https://www.heartflow.com/newsroom/heartflow-analysis-is-first-ai-enabled-technology-to-be-recognized-by-the-acc-and-aha-guidelines-as-important-tool-for-diagnosing-and-treating-heart-disease/
43. EchoNous, Point of Care Ultrasound Technology, URL: https://echonous.com/
44. The world's most advanced 3D anatomy platform, URL: https://3d4medical.com/
45. Babylon Health, https://www.emed.com/uk
46. Ada Health, https://ada.com/

Chapter 3
Principles of Machine Learning

3.1 Introduction

The fundamental principles of machine learning revolve around utilizing algorithms to enable computers to learn from data and make predictions or decisions without explicit programming. In the context of medical imaging and related fields, machine learning plays a crucial role in extracting meaningful insights from large volumes of medical images, such as X-rays, MRIs, CT scans, and microscopy images. It involves tasks such as feature extraction, pattern recognition, and classification.

These aspects enable the algorithms to identify relevant patterns or abnormalities within the medical images, leading to more accurate and efficient diagnosis and treatment. Additionally, the core elements of machine learning emphasize the importance of model training, validation, and interpretation to ensure the reliability and generalizability of the algorithms. Overall, these insights contribute to the advancement of medical diagnostics and personalized healthcare.

There are **four appendices** in this chapter. The primary aim of these appendices is to offer additional details on specific technical aspects, aiding in the comprehension of the workings of machine learning.

3.2 Overview of Subfields

Machine learning encompasses various subfields, each serving distinct purposes in the analysis of data. These subfields consist of supervised learning, unsupervised learning, and reinforcement learning [1].

Supervised learning involves training a model on labeled data, where the algorithm learns to map input data to the desired output [2]. In medical imaging, supervised learning is crucial for tasks such as image classification (identifying a specific

E. Seeram, V. Kanade, *Artificial Intelligence in Medical Imaging Technology*,
https://doi.org/10.1007/978-3-031-64049-0_3

condition from an image) and object detection (locating and categorizing specific features within an image), enabling the development of tools for automated detection of anomalies and diseases in medical images.

Unsupervised learning, on the other hand, deals with unlabeled data, seeking to uncover hidden patterns or structures within the data [3]. In medical imaging, unsupervised learning is relevant for tasks such as image segmentation (partitioning an image into meaningful regions), clustering (grouping similar images or features), and anomaly detection (identifying outliers or abnormalities), assisting in the exploration and understanding of complex medical image datasets.

Reinforcement learning involves an agent learning to make decisions by interacting with an environment to achieve a specific goal [4]. Although less prevalent in medical imaging, reinforcement learning holds potential in areas such as image-guided interventions and personalized treatment planning, where algorithms can learn to make sequential decisions in response to dynamic and complex imaging data. This ultimately optimizes patient care and outcomes. As medical imaging generates vast amounts of data, these subfields play a critical role in harnessing and interpreting this data for enhanced analysis, diagnosis, and treatment strategies.

3.2.1 Supervised Learning

Supervised learning, one of the predominant forms of machine learning, operates on the principle of utilizing labeled datasets to train algorithms to classify data or predict outcomes accurately. Labeled datasets are crucial as they provide both the input data and the desired output, often annotated by human experts, against which the model's accuracy can be measured and improved.

In technical terms, supervised learning algorithms create a mathematical model of a set of data that contains both the inputs and the desired outputs. The data is split into a training set and a test set, with the majority often allocated for training. The algorithm iteratively makes predictions on the training data and is corrected by the training algorithm when those predictions differ from the desired outcome. This training process adjusts the model's parameters, typically the weights of connections between nodes in the case of neural networks, until the model's predictions align closely with the actual outcomes.

Once the model has been trained to a satisfactory level of accuracy on the training data, it is then validated using the test set, which it has not seen before, to ensure that it can generalize well to new data (Fig. 3.1).

In medical imaging, supervised learning techniques provide an automated means to interpret vast amounts of imaging data that would be impractical for human experts to analyze individually. Here's how supervised learning is utilized within this field:

An example of detection of diabetic retinopathy in retinal images is provided in Table 3.1.

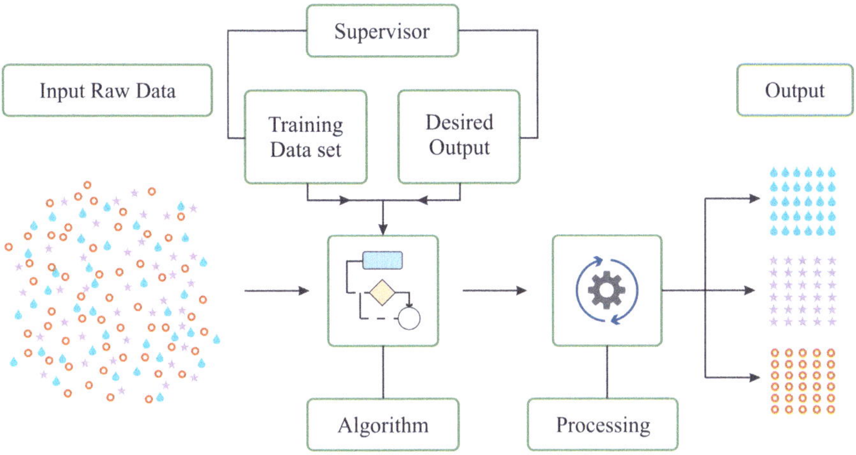

Fig. 3.1 Supervised learning technique. See text for further explanation

Table 3.1 Employing supervised learning to categorize retinal images as either indicative of diabetic retinopathy or not displaying such indications. Diabetic retinopathy is a condition that can cause vision loss in people with diabetes

Essential steps	Brief explanation
Data collection	A labeled dataset is gathered which consists of thousands of retinal images. Each image is labeled as showing signs of diabetic retinopathy or not, often with gradations indicating the severity of the disease
Preprocessing	The images are preprocessed to enhance certain features, reduce noise, and standardize the input for the learning algorithm. This might involve cropping images to the region of interest, normalizing pixel values, and using image augmentation techniques to increase the dataset size
Training phase	A supervised learning algorithm, for instance, a CNN, which is well-suited for visual data, is trained on this dataset. During training, the CNN uses filters to capture various aspects of the images, such as edges, textures, and patterns, and learns to differentiate between images with and without diabetic retinopathy
Validation and testing	The trained model is then validated and tested on a separate set of images not used during training. The performance of the model is assessed using metrics such as accuracy, precision, recall, and the area under the curve-receiver operating characteristic (AUC-ROC) curve
Deployment	Once the model meets the necessary accuracy thresholds, it can be deployed in clinical settings. It analyzes the retinal images of patients and predicts whether or not the signs of diabetic retinopathy are present
Clinical decision support	The model serves as a diagnostic aid for ophthalmologists, helping to prioritize cases and providing a second opinion. The final diagnosis would likely still be made by a medical professional, taking into account the model's findings, the patient's medical history, and other diagnostic tests

The strength of using supervised learning in medical imaging lies in its ability to handle large-scale data, its repeatability, and the consistency of its analyses. It reduces human error and increases the efficiency of the medical diagnostic process. Importantly, as diagnostic imaging continues to grow in volume and complexity, such AI-enhanced tools become ever more valuable for patient care and treatment planning [5].

Different supervised machine learning techniques and their applications in medical imaging and analysis are described in Appendix 1. The discussed techniques encompass linear regression, decision trees, and support vector machines (SVMs).

3.2.2 Unsupervised Learning

Unsupervised learning is a machine learning paradigm in which the algorithm is trained on unlabeled data to discover patterns, structures, or relationships within the data. Unlike supervised learning, there are no predefined target labels, and the algorithm's task is to explore the inherent structure of the data without explicit guidance (Fig. 3.2).

Clustering Algorithms

Clustering algorithms are a fundamental component of unsupervised learning [6]. They group data points into clusters based on their similarities, aiming to maximize intra-cluster similarity and minimize inter-cluster similarity. Clustering algorithms help uncover natural groupings or patterns within the data, which can be highly beneficial in medical image analysis for tasks such as disease subtyping, anomaly detection, or tissue segmentation.

Appendix 2 provides explanations for commonly used clustering algorithms.

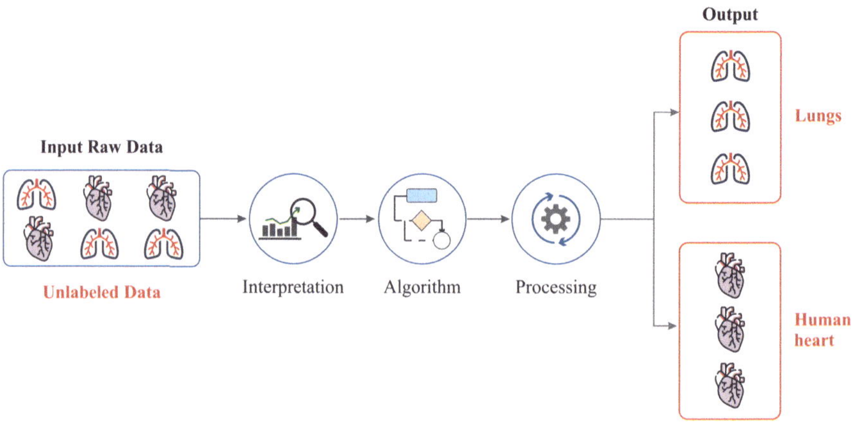

Fig. 3.2 Unsupervised learning method. See text for further explanation

3.2.3 Reinforcement Learning

Reinforcement learning (RL) is a subfield of machine learning that enables computers to learn autonomously by making a series of decisions and learning from the consequences of those decisions. Through trial and error, computer programs gradually refine their performance by receiving feedback, either positive or negative, based on their actions.

The core idea of RL is to learn the most advantageous behavior within a specific environment to maximize rewards, making it particularly suitable for problems involving interconnected decisions. This approach is beneficial in addressing complex challenges such as navigating walking robots or autonomous vehicles, where a sequence of decisions is necessary for optimal performance [7].

Unlike other learning frameworks that rely on preexisting datasets, RL acquires data through direct interaction with the environment. This unique characteristic allows the software to explore, engage with, and learn from its surroundings, discovering optimal solutions in the process.

In medical imaging, RL has been increasingly applied to tackle complex problems, and one important application is optimizing radiation dose in CT scans.

Key Reinforcement Learning Elements

- **Agent**: The entity that makes decisions based on the information it receives from the environment.
- **Environment**: The external system with which the agent interacts. It provides feedback and consequences to the agent's actions.
- **State**: The representation of the current situation, typically provided by the environment.
- **Action**: The decision or choice made by the agent that can influence the environment.
- **Reward**: A numerical value provided by the environment as feedback to the agent to evaluate the action taken. The goal is to maximize cumulative rewards.
- **Policy**: The strategy or plan that the agent follows to select actions in different states (Fig. 3.3).

Application: Optimizing Radiation Dose in CT Imaging

Medical images, especially CT scans, are essential for diagnosing a wide range of health conditions. However, excessive radiation exposure from CT scans can be harmful. The challenge is to reduce radiation dose while maintaining the image quality necessary for accurate diagnosis.

In this context, reinforcement learning can be applied as follows:

- **Environment**: The CT scanner and the patient's body form the environment. The state represents the current situation of the scan, including the position of the CT scanner, the type of scan (e.g., head, chest, or abdomen), and the patient's physical characteristics.
- **Agent**: The agent could be the CT scanner itself or a computer system controlling the scanner's parameters.

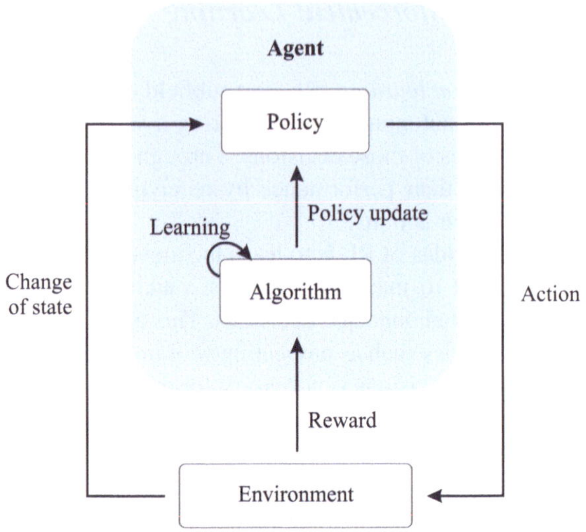

Fig. 3.3 Reinforcement learning elements. See text for further explanation

- **Action**: The agent selects actions related to CT scan parameters, such as tube current, tube voltage, and scan duration. These parameters directly affect the radiation dose delivered to the patient.
- **Reward**: The reward in this scenario can be a combination of factors. It includes image quality (high-quality images receive higher rewards), radiation dose (lower doses receive higher rewards), and patient safety. The reward should encourage the agent to find the right balance between minimizing radiation and maintaining image quality.
- **Policy**: The agent uses a policy to choose actions in different situations. This policy is learned through trial and error, as the agent explores different parameter settings and their outcomes.

Example

Consider a scenario where a hospital possesses a CT scanner, and the objective is to minimize radiation dose without compromising the quality of the images. The RL agent begins with a default policy. It starts by selecting different combinations of tube current and tube voltage for different types of scans. The agent then scans patients and receives feedback in the form of image quality scores and the radiation dose delivered.

Over time, the agent learns which parameter settings result in high-quality images with the least radiation dose. It adapts its policy accordingly. For instance, the agent might learn that for a head CT scan on a pediatric patient, a lower tube voltage and shorter scan duration produce acceptable images with less radiation. It then applies this knowledge to future scans.

Reinforcement learning allows the agent to adapt to different patient profiles and scan types. It continuously refines its policy to optimize radiation dose while maintaining diagnostic image quality, making it a valuable tool for improving patient safety in medical imaging.

3.3 Key Components of a Machine Learning System

In the rapidly evolving landscape of medical technology, the integration of machine learning has become pivotal for advancing overall diagnosis. Within this intricate realm, understanding the fundamental components of a machine learning system is paramount. This section illuminates the intricate machinery underpinning the seamless fusion of artificial intelligence and healthcare. From robust data acquisition mechanisms to sophisticated algorithms designed to analyze vast datasets, the exploration delves into the critical elements that empower medical professionals to harness the potential of machine learning (Fig. 3.4).

3.3.1 Data Acquisition and Preparation

3.3.1.1 Data Acquisition

Data acquisition entails collecting and gathering the necessary data for a machine learning project. This typically involves various types of data obtained from different sources, such as patient records, medical images, laboratory results, and more. Here are some key points to consider:

Data Sources

- **Electronic health records (EHR)**: These records contain patient demographics, medical history, diagnoses, medications, and other clinical information. EHRs aim to provide a centralized and digital repository of a patient's health information for efficient and comprehensive healthcare management.

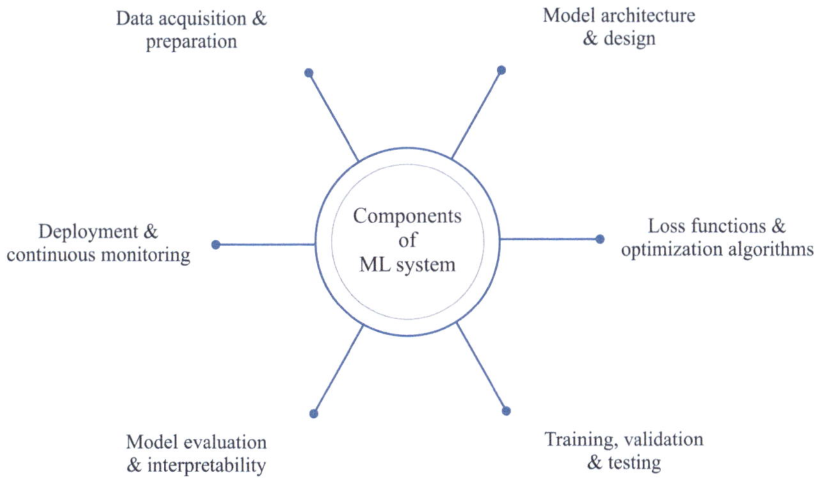

Fig. 3.4 Key components of the ML system. See text for further explanation

- **Medical imaging**: This data includes X-rays, MRIs, CT scans, and histopathology images. For instance, a radiology department can provide a dataset of chest X-rays for a pneumonia detection model.
- **Wearable devices**: Devices such as heart rate monitors, fitness trackers, and smartwatches collect patient data over time. This information can be valuable for tracking trends in patient health.
- **Clinical trials**: This involves information gathered through clinical trials, encompassing how patients react to treatments and medications. Such data can even be utilized in research and the development of pharmaceuticals.

Data Privacy and Compliance

In the realm of healthcare, safeguarding *data* privacy is of paramount importance, necessitating meticulous adherence to stringent regulatory frameworks. Particularly in the United States, the Health Insurance Portability and Accountability Act (HIPAA) stands as a cornerstone in the protection of patient information [8]. This legislation mandates strict protocols for healthcare professionals, compelling them to implement robust security measures to safeguard sensitive medical data.

HIPAA not only emphasizes the confidentiality of patient records but also places a significant emphasis on ensuring the integrity and accessibility of health information. Compliance with such regulatory requirements is indispensable to foster trust among patients, guaranteeing that their personal and medical information remains confidential and secure.

The significance of data protection laws extends beyond the United States, with many countries globally implementing their own comprehensive frameworks. These laws are designed to address the evolving challenges posed by the digital age and ensure that medical institutions adopt practices that prioritize the privacy and security of patient information. Compliance with these laws not only shields healthcare organizations from legal ramifications but also underscores their commitment to ethical and responsible data handling.

Hence, medical professionals must ensure that they adhere to regulatory requirements such as the HIPAA and abide by the compliance rules or data protection guidelines.

Data Cleaning and Preprocessing

Raw data from data sources often contain errors, missing values, and inconsistencies. For example, patient records may have typos or missing information. It is crucial to clean and preprocess the data to make it suitable for machine learning processes. This might involve removing duplicates, imputing missing values, and standardizing formats.

3.3.1.2 Data Preparation

Once the data is acquired, it needs to be prepared for use in machine learning models. This involves various tasks:

(a) **Feature Engineering**

In medical applications, features encompass patient demographics, vital signs, lab results, or specific findings from medical images. Feature engineering encompasses the selection, transformation, or creation of new features relevant to the problem being addressed. For instance, a patient's body mass index (BMI) could be calculated from their height and weight as part of this process.

(b) **Data Splitting**

Typically, the data is divided into training, validation, and test sets. The training set is used to train the machine learning model, the validation set is used for hyperparameter tuning and model evaluation, and the test set is used to assess the model's generalization to new, unseen data.

(c) **Data Balancing**

In medical datasets, class imbalance is a common occurrence, particularly when detecting rare diseases where there may be a substantial disparity between negative and positive cases, with a higher prevalence of negative instances. To train a fair and accurate model, data balancing techniques, including oversampling, undersampling, or generating synthetic data, may be necessary.

(d) **Data Normalization and Scaling**

Various features in a dataset may have different scales; for example, blood pressure readings are measured in millimeters of mercury (mmHg), while age is measured in years. Scaling or normalizing these features can aid in enhancing model training and convergence.

(e) **Data Augmentation (for Image Data)**

Data augmentation, particularly relevant for image data, is a process used to enhance the diversity of the dataset for training machine learning models. By applying random geometric transformations such as rotations, flips, scaling, or cropping to the original images, data augmentation artificially generates new training examples. This increased diversity helps the model generalize better and become more robust to variations and noise in the real-world data.

In medical imaging analysis, where labeled data is often limited and valuable, data augmentation can be crucial for improving model performance by effectively expanding the training dataset without the need for additional manual annotation. For instance, it can help a model recognize anatomical structures or abnormalities from various angles and orientations, making it more effective in clinical applications.

(f) **Handling Time Series Data**

Handling time series data is crucial in various medical applications, particularly those involving continuous vital sign monitoring. In such scenarios, where data is collected over time, specialized approaches become essential for effective analysis.

Techniques such as recurrent neural networks (RNNs) or long short-term memory (LSTM) networks are often employed to address the temporal nature of the data [9, 10]. These models excel at capturing sequential dependencies and patterns within the time series, making them well-suited for tasks that involve predicting or understanding trends in medical data over extended periods.

Data acquisition and preparation are critical steps in building machine learning models for medical professionals. Properly sourced and prepared data can significantly impact the accuracy and effectiveness of models used in tasks such as disease diagnosis, patient monitoring, and drug discovery. Moreover, the ethical handling of patient data and adherence to privacy regulations are of utmost importance in medical machine learning applications.

3.3.2 Model Architecture and Design

Model architecture and design is another key component of a machine learning system, particularly for medical professionals who seek to develop predictive or diagnostic models. In this context, it involves designing the structure and characteristics of the machine learning model that will ultimately process and analyze medical data.

3.3.2.1 Model Selection

Choosing the right machine learning model architecture is crucial. The choice of model depends on the specific task and the nature of the data. Here are a few examples:

- **Logistic regression**: It's a simple and interpretable model often used for binary classification tasks, such as predicting whether a patient has a particular disease or not.
- **Convolutional neural networks (CNNs)**: These are commonly used for medical image analysis. For instance, a CNN model can be designed to detect tumors in medical images such as mammograms or MRIs.
- **Recurrent neural networks (RNNs) or long short-term memory (LSTM) networks**: These are suitable for analyzing time series data, making them ideal for tasks such as predicting patient outcomes over time, monitoring vital signs, or analyzing electroencephalogram (EEG) data.
- **Transformer-based models**: These models, such as bidirectional encoder representations from transformers (BERT) or generative pretrained transformer (GPT), are being increasingly used in natural language processing tasks in healthcare, such as medical record summarization or clinical text analysis [11].

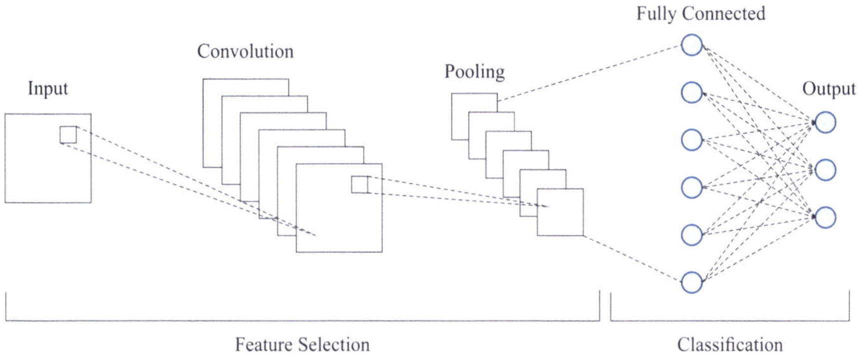

Fig. 3.5 CNN for medical image analysis. See text for further explanation

3.3.2.2 Model Architecture

Once a model type has been selected, it becomes necessary to design its architecture. Typically, the chosen model type serves as a foundational framework, but the subsequent step involves crafting a specific architecture tailored to the task at hand. The architecture essentially outlines the blueprint of the model, specifying critical components such as the number and size of layers, as well as the intricate connections between them. These decisions significantly influence the model's ability to learn and generalize from the provided data.

For instance, consider a scenario where a CNN is employed for medical image analysis. In this case, the architecture might feature multiple convolutional layers, designed to extract intricate patterns from the input images. Subsequently, pooling layers could be incorporated to downsample the spatial dimensions and capture essential hierarchical information. Finally, fully connected layers may follow, facilitating the classification of the extracted features into specific medical categories. The thoughtful design of this architecture enables the CNN to effectively process and understand complex visual information, demonstrating the importance of architectural choices in optimizing model performance for specific tasks (Fig. 3.5).

3.3.2.3 Hyperparameter Tuning

Model design also involves tuning hyperparameters, which are settings that are not learned from the data but have a significant impact on the model's performance [12]. For example:

- **Learning rate**: This hyperparameter controls how quickly the model adjusts its internal parameters during training. It needs to be carefully chosen to ensure convergence and prevent overshooting or slow learning.

- **Number of layers and units**: The depth and width of the neural network, for instance, in a feedforward neural network, should be determined based on the complexity of the problem. Deep networks can capture intricate patterns but may also be prone to overfitting.

3.3.2.4 Regularization and Optimization

Regularization techniques are essential in preventing overfitting, where a model becomes too closely tailored to the training data and loses its ability to generalize to new instances. In medical imaging analysis, this is crucial for ensuring a model's robustness and accuracy in real-world scenarios. Common regularization techniques are the following:

- **L1 and L2 regularization**: This method adds penalty terms to the loss function based on model weight magnitudes. For instance, L2 regularization discourages large weight values, aiding in preventing overfitting and directing the model to focus on crucial features [13].

Imagine you have a mathematical model that tries to learn from data to make predictions, such as predicting the price of a house based on its features. Now, this model has some "weights" assigned to different features (such as the number of bedrooms and square footage).

L1 and L2 regularization: These are methods to prevent the model from becoming too complex and overfitting the data. Overfitting is when the model learns the training data too well and performs poorly on new, unseen data.

Penalty terms: L1 and L2 regularization add extra terms to the mathematical formula the model uses to make predictions. These extra terms act as penalties.

Model weight magnitudes: The "weights" in the model represent the importance given to different features. L1 and L2 regularization penalize these weights based on their magnitudes (how big or small they are).

L2 regularization example: Specifically, L2 regularization discourages the weights from becoming too large. It does this by adding a penalty to the loss function (a measure of how well the model is doing) that depends on the squares of the weights. This helps to keep the weights from growing too much, which can be beneficial in preventing overfitting.

In simple terms, L2 regularization helps keep the model from getting too obsessed with specific details in the training data and encourages it to focus on the essential information for making accurate predictions on new data.

- **Dropout**: Dropout is another technique that randomly deactivates a fraction of neurons during training in a neural network. This prevents overreliance on specific neurons, promoting the learning of more robust features [13]. In medical image classification, dropout ensures that no single pixel or feature dominates decision-making.

- **Batch normalization**: Batch normalization normalizes neuron activations within a layer, enhancing training stability. In medical imaging tasks, it facilitates faster convergence and improves generalization by reducing internal covariate shift.
- Internal covariate shift refers to the change in the distribution of the input to a neural network's layers during training. In the context of neural networks, each layer's input distribution can change as the parameters of the previous layers are updated during training. This can make the training process more challenging because the network has to continuously adapt to these changing distributions. Hence, batch normalization technique is employed to address internal covariate shift.

For example, in a radiology setting, training a deep convolutional neural network for anomaly detection in X-ray images involves applying L2 regularization to convolutional layers to prevent overfitting and dropout to fully connected layers for robustness.

3.3.2.5 Pretrained Models and Transfer Learning

Transfer learning entails utilizing the knowledge and features of a pre-trained model on a related task to enhance the performance of a model on a new but associated task. In medical imaging, this approach is highly effective, especially considering the challenges associated with collecting large medical image datasets and addressing complexities of image analysis.

Example Consider a hospital that wants to develop a model to detect diabetic retinopathy from retinal images. They can take a pre-trained convolutional neural network such as VGG16 or ResNet, which has been trained on a large dataset such as ImageNet, and fine-tune it on their diabetic retinopathy dataset [14]. This is advantageous because the lower layers of these networks have already learned basic features such as edges and textures, which are relevant to detecting abnormalities in retinal images. Fine-tuning on the specific task of diabetic retinopathy will adjust the higher-level features to be more relevant to the medical domain.

Thus, pretrained models and transfer learning save time and computational resources while improving model performance, making them a valuable technique in medical imaging analysis.

3.3.2.6 Interpretability and Explainability

In the field of medical imaging, model interpretability is crucial for several reasons, including trust, clinical adoption, and regulatory compliance. Healthcare professionals and patients need to understand why a model makes a specific diagnosis or prediction. Here are some techniques and their examples:

- **SHapley Additive exPlanations (SHAP) values**: SHAP values provide a way to explain the output of any machine learning model by attributing the contribution of each feature to the final prediction [15]. For instance, in the analysis of abdominal CT scans for liver disease classification, SHAP values can elucidate which regions or features within the scans played a significant role in the model's prediction. This can offer valuable insights into the decision-making process and enhance confidence in the model's accuracy.
- **Local interpretable model-agnostic explanations (LIME)**: LIME is a technique that generates local explanations for individual predictions [16]. In the context of cardiology, where an algorithm is tasked with identifying abnormalities in echocardiogram images, LIME can create a simplified, interpretable model for a specific image, shedding light on the features that contributed to the diagnostic decision.
- **Saliency maps**: Saliency maps highlight the most important regions of an image with respect to the model's prediction [17]. For instance, in musculoskeletal imaging, a model focused on detecting abnormalities in X-rays of joints can generate saliency maps to reveal which areas in the X-ray influenced its decision. This can assist orthopedic specialists in comprehending the model's reasoning and decision-making process.

These techniques offer transparency and insights into the model's decision-making process, making it easier for healthcare professionals to trust and use AI-powered tools effectively.

3.3.2.7 Validation and Evaluation

Model validation and evaluation are critical to ensure that medical imaging models perform as expected and align with clinical objectives. Here are some considerations and examples:

- **Cross-validation**: Cross-validation is a technique used to assess a model's generalization performance by splitting the dataset into multiple subsets, training the model on different subsets, and evaluating it on the remaining data. In a mammography analysis task, a fivefold cross-validation can help determine how well the model generalizes across different subsets of mammogram images, reducing the risk of overfitting.
- **Appropriate metrics**: The choice of evaluation metrics is essential and depends on the specific clinical problem. For instance, consider a model designed for predicting osteoporosis risk based on bone density scans. In this scenario, metrics such as positive predictive value, negative predictive value, accuracy, and Matthew's correlation coefficient could be employed to assess the model's performance [18]. These metrics offer insights into the model's ability to correctly identify individuals at risk of osteoporosis while minimizing both false positives and false negatives.

Rigorous validation and evaluation help ensure that the model is both accurate and reliable for clinical use, meeting the standards of medical practice and regulatory requirements. Thus, interpretability and explainability techniques enable medical professionals to understand and trust AI models' decisions.

3.3.3 Loss Functions and Optimization Algorithms

"Loss functions and optimization algorithms" are essential tools in machine learning systems that fine-tune models, optimize their accuracy, and ensure effective performance in diagnostic applications by minimizing errors. These concepts have been described in detail in this section to explore how they relate to medical practice:

3.3.3.1 Loss Functions

A loss function, also known as a cost function, serves as a critical component for measuring the performance of machine learning models. It quantifies the gap between the model's predictions and the actual clinical data. In essence, it evaluates how well the model is doing its job. Different medical tasks require specific loss functions, each with a distinct purpose:

- **Mean squared error (MSE)**: Used in tasks such as predicting patient age or blood pressure, where precise numerical predictions are desired. It calculates the squared difference between the model's predictions and the actual values, helping us gauge how close the model's predictions are to the real measurements.
- **Binary cross-entropy**: Applied in binary classification tasks, such as detecting the presence or absence of a disease, such as COVID-19 [19]. It quantifies the difference between the model's probability estimates and the actual binary outcomes, helping us assess the model's ability to classify correctly.
- **Categorical cross-entropy**: Essential in multiclass classification tasks, such as classifying diseases into different categories or types. It measures how well the model's probability distribution across different classes aligns with the actual class labels, enabling accurate classification of various conditions.
- **Dice coefficient**: Commonly used in tasks such as tumor segmentation within medical images, such as MRI scans. It assesses the overlap between the model's predicted segmentation of a tumor and the actual segmentation, ensuring precise identification of the tumor's boundaries in images.

Thus, selecting the appropriate loss function is a critical step in training machine learning models, as it guides the model to learn and improve its predictive accuracy, ultimately contributing to more accurate diagnosis, prognosis, and treatment recommendations for patients.

3.3.3.2 Optimization Algorithms

Optimization algorithms are used to adjust the model's parameters (weights and biases) during training to minimize the loss function. Different optimization algorithms have their strengths and are suitable for various machine learning models. These algorithms include stochastic gradient descent (SGD), adaptive moment estimation (Adam), root mean square propagation (RMSprop), adaptive gradient descent (AdaGrad), Limited-memory Broyden-Fletcher-Goldfarb-Shanno (LBFGS), and conjugate gradient [20–22].

For example, in the diagnosis of tumors from medical images, SGD would adjust the model parameters by considering small, random batches of patient images in each iteration, gradually improving the model's ability to accurately identify and classify tumors while efficiently navigating through large datasets. A detailed explanation of the optimization algorithms is presented in Appendix 3.

Selecting the appropriate optimization algorithm depends on factors such as the problem at hand; the model architecture; characteristics of medical imaging data, such as noise and variability; and the dataset size. It's common to experiment with multiple optimizers to find the one that converges quickly and effectively for a specific task.

In the context of healthcare, choosing the right loss function and optimization algorithm is critical for building accurate and reliable machine learning models.

3.3.4 Training, Validation, and Testing

The processes of *training*, *validation*, and *testing* are fundamental elements within a machine learning system, playing a pivotal role in the development and assessment of machine learning models. This section throws light on these critical aspects along with examples relevant to the healthcare domain.

3.3.4.1 Training

In the training phase, the machine learning model learns from a labeled dataset (input-output pairs) to understand the underlying patterns and relationships within the data. The model adjusts its parameters to minimize a defined loss function. Training is an iterative process, and the model continues to improve its performance as it learns from the data.

Example Consider developing a machine learning model for the early detection of pneumonia using chest X-ray images. The model is trained on a dataset containing chest X-ray images, each labeled as either healthy or displaying early signs of pneumonia. Through the training process, the model learns to identify relevant features and patterns that indicate the presence of the respiratory condition.

3.3.4.2 Validation

The validation phase is used to fine-tune the model's hyperparameters and to assess its performance on a dataset that it hasn't seen during training. This helps in preventing overfitting (where the model performs well on the training data but poorly on unseen data) and ensures that the model generalizes well to new, unseen use cases.

Example Consider developing a model for predicting patient outcomes in a critical care unit. During validation, hyperparameters such as the learning rate, the number of layers in the neural network, or the batch size are adjusted to optimize the model's performance on a validation dataset. This ensures that the model doesn't overfit to the training data and is robust when applied to new patient cases.

3.3.4.3 Testing

The testing phase evaluates the model's generalization ability and assesses its performance on completely new data. The testing dataset is separate from both the training and validation data. It provides an estimate of how well the model will perform in real-world, clinical applications.

Example To assess the diagnostic performance of a machine learning model for breast cancer detection using mammograms, a separate testing dataset would be employed, consisting of mammograms not utilized in the training or validation phases. The model's capacity to accurately distinguish between cancerous and non-cancerous cases in this dataset serves as an indicator of its real-world applicability.

In summary, the training, validation, and testing phases are integral to the development and assessment of machine learning models in healthcare. The entire process is revealed in Fig. 3.6.

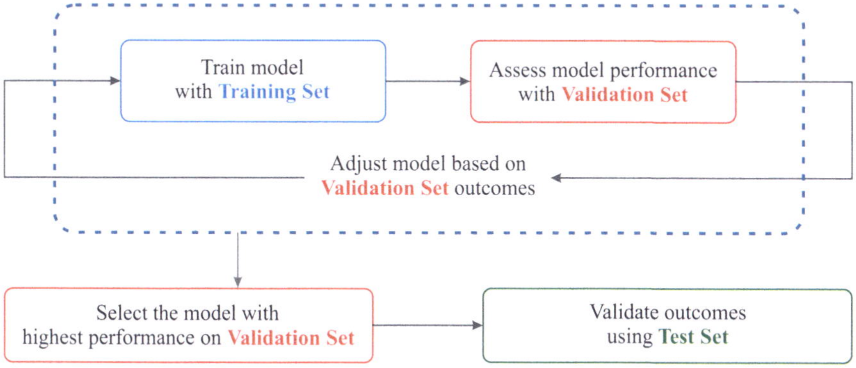

Fig. 3.6 Training, validation, and testing phases. See text for further explanation

3.3.5 Model Evaluation and Interpretability

Understanding and assessing the performance of a machine learning model, as well as its interpretability, are crucial aspects of a machine learning system. Let's understand these components in detail with examples relevant to medical applications:

3.3.5.1 Model Evaluation

Model evaluation involves assessing the performance and reliability of a machine learning model by quantifying its ability to make accurate predictions. In healthcare, model evaluation is vital because the consequences of erroneous predictions can have a significant impact on patient outcomes.

Example Consider a scenario where a machine learning model has been developed to predict whether a patient has a rare disease based on their clinical data. To evaluate the model, a dataset of patient cases with known disease outcomes is used. The model's performance can be assessed using metrics such as sensitivity (true positive rate), specificity (true negative rate), accuracy, and the receiver operating characteristic (ROC) curve.

3.3.5.2 Model Interpretability

Model interpretability refers to the capacity to understand, explain, and interpret the reasoning behind a model's predictions. Interpretability is crucial because healthcare professionals need to trust the model's output and comprehend the factors influencing a particular prediction.

Example Consider a scenario where a machine learning model has been constructed to assist physicians in selecting the most appropriate treatment for patients with a specific condition. Interpretability enables the explanation of why the model recommends a particular treatment, highlighting the features (clinical parameters) that had the most significant impact on the decision. This transparency aids healthcare providers in understanding and trusting the model's recommendations.

Hence, model evaluation and interpretability are vital components in a machine learning system. These components ensure that machine learning models are not only accurate but also understandable and ethically sound, making them valuable tools for improving patient care and clinical decision-making.

3.3.6 *Deployment and Continuous Monitoring*

"Deployment and continuous monitoring" involves taking a trained machine learning model from the development phase and integrating it into the clinical environment, as well as continuously monitoring its performance and safety. Here's a detailed explanation with examples for medical professionals:

3.3.6.1 Deployment

Deployment is the process of integrating a machine learning model into the clinical workflow, making it accessible for use in a real-world setting.

Example Consider a scenario where a machine learning model has been developed to predict the risk of stroke in patients based on their medical history and clinical data. Deployment would involve integrating this model into an electronic health record (EHR) system. Consequently, when a physician accesses a patient's record, they can also view the predicted stroke risk, aiding in clinical decision-making.

3.3.6.2 Continuous Monitoring

Continuous monitoring involves tracking the model's performance in a clinical setting over time. It ensures that the model remains accurate, safe, and relevant as data distributions, patient populations, or medical practices evolve.

Example Consider a machine learning model deployed for identifying abnormalities in medical images, such as chest X-rays. Continuous monitoring would involve regularly assessing the model's performance, its ability to detect new types of abnormalities, and its performance on different patient demographics.

Thus, the deployment and continuous monitoring of machine learning models in healthcare are critical for their real-world utility and safety. By integrating models into clinical workflows and continually assessing their performance and relevance, medical professionals can rely on these tools to support clinical decision-making, diagnosis, and treatment planning while ensuring data privacy and safety.

3.4 Evaluation Metrics for Machine Learning Models

Evaluation metrics for machine learning models play a critical role as they provide a quantitative measure of a model's performance, helping researchers and healthcare professionals assess the model's accuracy, reliability, and clinical applicability.

These metrics include classification metrics, ROC curve and AUC-ROC, precision-recall curve, confusion matrix analysis, intersection over union (IoU), and dice coefficient [23–26]. In medical imaging, where the consequences of incorrect diagnoses can be life-altering, these metrics are of utmost importance.

Common evaluation metrics used in this field include sensitivity, specificity, accuracy, and the area under the receiver operating characteristic curve (AUC-ROC). Sensitivity measures the ability of a model to correctly identify positive cases (e.g., disease presence), while specificity quantifies the model's ability to correctly identify negative cases (e.g., disease absence). High sensitivity is crucial to avoid false negatives, ensuring diseases are not missed, while high specificity helps minimize false positives, reducing unnecessary further tests and anxiety for patients. Accuracy gives an overall measure of a model's correctness, and AUC-ROC provides a comprehensive assessment of a model's ability to distinguish between different classes, which is particularly relevant in binary classification tasks such as tumor detection.

Furthermore, evaluation metrics play a pivotal role in model selection and optimization. Medical imaging analysis often involves a trade-off between sensitivity and specificity, where increasing one may decrease the other. Evaluation metrics help strike the right balance by allowing researchers to select models that align with the specific clinical requirements and the relative importance of avoiding false negatives or false positives in a given medical application. Additionally, they enable the fine-tuning of model parameters and architectures to achieve optimal performance. This ensures that the model meets the high standards of accuracy and reliability demanded by medical professionals, making it a crucial aspect of developing machine learning solutions for imaging analysis.

A further elaboration of these evaluation metrics is given in Appendix 4.

References

1. Sara Brown, Machine learning, explained, Apr 21, 2021, URL: https://mitsloan.mit.edu/ideas-made-to-matter/machine-learning-explained
2. IBM, What is supervised learning? URL: https://www.ibm.com/topics/supervised-learning
3. IBM, What is unsupervised learning? URL: https://www.ibm.com/topics/unsupervised-learning
4. Bhatnagar, S., Prasad, H., Prashanth, L. (2013). Reinforcement Learning. In: Stochastic Recursive Algorithms for Optimization. Lecture Notes in Control and Information Sciences, vol 434. Springer, London. https://doi.org/10.1007/978-1-4471-4285-0_11
5. Open Medscience, Revolutionising Medical Imaging with AI and Big Data Analytics, URL: https://openmedscience.com/revolutionising-medical-imaging-with-ai-and-big-data-analytics/#:~:text=AI%20can%20help%20researchers%20develop,and%20develop%20personalised%20treatment%20plans.
6. Teoh, T.T., Rong, Z. (2022). Clustering. In: Artificial Intelligence with Python. Machine Learning: Foundations, Methodologies, and Applications. Springer, Singapore. https://doi.org/10.1007/978-981-16-8615-3_12
7. Escobar-Naranjo, Juan, Gustavo Caiza, Paulina Ayala, Edisson Jordan, Carlos A. Garcia, and Marcelo V. Garcia. 2023. "Autonomous Navigation of Robots: Optimization with DQN" Applied Sciences 13, no. 12: 7202. https://doi.org/10.3390/app13127202

8. CDC, Health Insurance Portability and Accountability Act of 1996 (HIPAA), URL: https://www.cdc.gov/phlp/publications/topic/hipaa.html#:~:text=The%20Health%20Insurance%20Portability%20and,the%20patient's%20consent%20or%20knowledge.
9. Das, S., Tariq, A., Santos, T., Kantareddy, S.S., Banerjee, I. (2023). Recurrent Neural Networks (RNNs): Architectures, Training Tricks, and Introduction to Influential Research. In: Colliot, O. (eds) Machine Learning for Brain Disorders. Neuromethods, vol 197. Humana, New York, NY. https://doi.org/10.1007/978-1-0716-3195-9_4
10. Zhang, J., Zeng, Y. & Starly, B. Recurrent neural networks with long term temporal dependencies in machine tool wear diagnosis and prognosis. SN Appl. Sci. 3, 442 (2021). https://doi.org/10.1007/s42452-021-04427-5
11. Shuang Zhang, Rui Fan, Yuti Liu, Shuang Chen, Qiao Liu, Wanwen Zeng, Applications of transformer-based language models in bioinformatics: a survey, Bioinformatics Advances, Volume 3, Issue 1, 2023, vbad001, https://doi.org/10.1093/bioadv/vbad001
12. Ippolito, P.P. (2022). Hyperparameter Tuning. In: Egger, R. (eds) Applied Data Science in Tourism. Tourism on the Verge. Springer, Cham. https://doi.org/10.1007/978-3-030-88389-8_12
13. Mehdi, C.A., Nour-Eddine, J., Mohamed, E. (2023). Regularization in CNN: A Mathematical Study for L1, L2, and Dropout Regularizers. In: Kacprzyk, J., Ezziyyani, M., Balas, V.E. (eds) International Conference on Advanced Intelligent Systems for Sustainable Development. AI2SD 2022. Lecture Notes in Networks and Systems, vol 637. Springer, Cham. https://doi.org/10.1007/978-3-031-26384-2_38
14. AbouEl-Magd, L.M., Darwish, A., Snasel, V. et al. A pre-trained convolutional neural network with optimized capsule networks for chest X-rays COVID-19 diagnosis. Cluster Comput 26, 1389–1403 (2023). https://doi.org/10.1007/s10586-022-03703-2
15. Rodríguez-Pérez, R., Bajorath, J. Interpretation of machine learning models using shapley values: application to compound potency and multi-target activity predictions. J Comput Aided Mol Des 34, 1013–1026 (2020). https://doi.org/10.1007/s10822-020-00314-0
16. Stadlhofer, A., Mezhuyev, V. Approach to provide interpretability in machine learning models for image classification. Industrial Artificial Intelligence 1, 10 (2023). https://doi.org/10.1007/s44244-023-00009-z
17. Bargal, S.A. et al. (2022). Beyond the Visual Analysis of Deep Model Saliency. In: Holzinger, A., Goebel, R., Fong, R., Moon, T., Müller, KR., Samek, W. (eds) xxAI – Beyond Explainable AI. xxAI 2020. Lecture Notes in Computer Science(), vol 13200. Springer, Cham. https://doi.org/10.1007/978-3-031-04083-2_13
18. Lim HK, Ha HI, Park SY, Han J. Prediction of femoral osteoporosis using machine-learning analysis with radiomics features and abdomen-pelvic CT: A retrospective single center preliminary study. PLoS One. 2021 Mar 4;16(3):e0247330. https://doi.org/10.1371/journal.pone.0247330. PMID: 33661911; PMCID: PMC7932154.
19. Hassan, E., Shams, M.Y., Hikal, N.A. et al. Detecting COVID-19 in chest CT images based on several pre-trained models. Multimed Tools Appl (2024). https://doi.org/10.1007/s11042-023-17990-3
20. Hassan, E., Shams, M.Y., Hikal, N.A. et al. The effect of choosing optimizer algorithms to improve computer vision tasks: a comparative study. Multimed Tools Appl 82, 16591–16633 (2023). https://doi.org/10.1007/s11042-022-13820-0
21. Andrei, N. (2008). Performance Profiles of Conjugate-Gradient Algorithms for Unconstrained Optimization. In: Floudas, C., Pardalos, P. (eds) Encyclopedia of Optimization. Springer, Boston, MA. https://doi.org/10.1007/978-0-387-74759-0_506
22. Liu, D.C., Nocedal, J. On the limited memory BFGS method for large scale optimization. Mathematical Programming 45, 503–528 (1989). https://doi.org/10.1007/BF01589116
23. Melo, F. (2013). Receiver Operating Characteristic (ROC) Curve. In: Dubitzky, W., Wolkenhauer, O., Cho, KH., Yokota, H. (eds) Encyclopedia of Systems Biology. Springer, New York, NY. https://doi.org/10.1007/978-1-4419-9863-7_242
24. Flach, P.A. (2011). ROC Analysis. In: Sammut, C., Webb, G.I. (eds) Encyclopedia of Machine Learning. Springer, Boston, MA. https://doi.org/10.1007/978-0-387-30164-8_733

25. Varoquaux, G., Colliot, O. (2023). Evaluating Machine Learning Models and Their Diagnostic Value. In: Colliot, O. (eds) Machine Learning for Brain Disorders. Neuromethods, vol 197. Humana, New York, NY. https://doi.org/10.1007/978-1-0716-3195-9_20
26. Gupta, M., Mishra, A. A systematic review of deep learning based image segmentation to detect polyp. Artif Intell Rev 57, 7 (2024). https://doi.org/10.1007/s10462-023-10621-1

Chapter 4
Principles of Deep Learning

4.1 Introduction to Neural Networks and Deep Learning Architectures

In the world of medical imaging, neural networks and deep learning are like supercharged tools that can help us understand images in a smarter way. Think of neural networks as intelligent assistants that learn from lots of examples to spot important details in medical scans, such as X-rays or MRIs. These networks are built like interconnected brain cells, where each cell helps recognize specific patterns, allowing them to quickly identify things such as tumors or abnormalities.

Deep learning takes this idea further—it's like giving these networks a power boost. By using deep learning, these networks become experts at recognizing complex patterns in images without needing tons of explicit instructions [1]. It's like training them to understand what a healthy lung looks like by showing them many examples, so they can spot irregularities or potential problems in new images they haven't seen before.

The architectures behind deep learning are like specialized blueprints guiding these networks. For example, convolutional networks are fantastic at handling medical images because they're designed to understand shapes and structures within these pictures. They help in tasks such as detecting diseases or tracking changes in a patient's condition over time. These architectures serve as powerful tools for medical imaging experts, helping them interpret scans more accurately and potentially detect issues at an earlier stage.

The following section explores the essential principles of neural networks and deep learning, specifically within the realm of medical imaging.

E. Seeram, V. Kanade, *Artificial Intelligence in Medical Imaging Technology*,
https://doi.org/10.1007/978-3-031-64049-0_4

4.2 Neural Network Fundamentals

Neural networks, empowered by backpropagation, refer to the basic building blocks of brain-inspired models used in understanding and processing data [2]. Imagine a neural network as a team of specialists working together to solve a problem, such as identifying diseases in medical images.

For instance, consider the task of diagnosing skin cancer using images of moles. Each "neuron" in the network corresponds to a specialist—some recognize edges and shapes, while others identify specific features such as color or texture. Backpropagation allows the network to learn from examples of both healthy and cancerous moles, adjusting the "knowledge" of each specialist (neuron) to become better at distinguishing between the two (Fig. 4.1).

Here's the kicker: when a new mole image is presented to the network, these specialists collaborate—passing information back and forth through the process known as backpropagation—until the network as a whole can confidently say whether the mole is indicative of cancer or not. The network's ability to learn from different examples and improve its accuracy over time, driven by backpropagation, is the essence of neural network fundamentals.

In a medical context, this principle is vital for tasks such as detecting anomalies in X-rays or MRIs. Just like a doctor who gets better at identifying diseases with experience, neural networks, use data to become more proficient at diagnosing conditions. This offers a powerful tool for medical professionals, enhancing accuracy and efficiency in healthcare diagnostics.

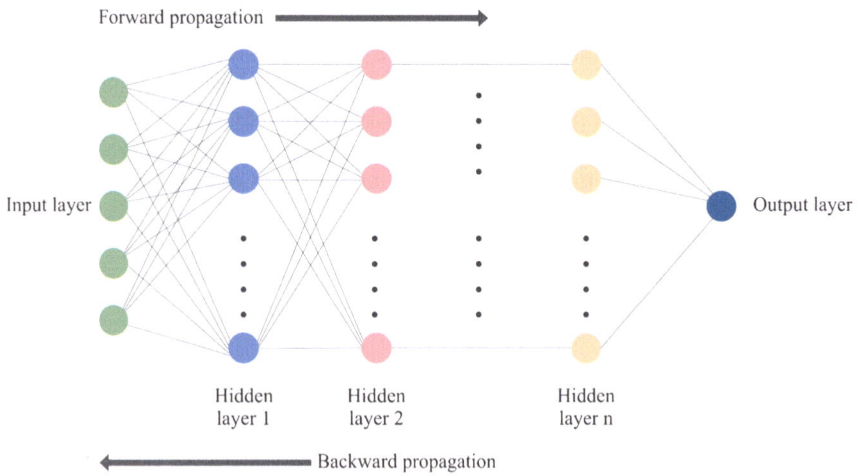

Fig. 4.1 A typical neural network. See text for further explanation

4.2.1 Feedforward Neural Networks

Feedforward neural networks are like organized pipelines that process information in one direction—from input to output—without any loops or cycles [3, 4]. Imagine a conveyer belt moving through different stations, with each station doing something specific to the item passing through. In medical imaging, think of a feedforward network as a team that looks at an X-ray of a patient's lungs and decides if there are signs of pneumonia or not.

Breaking it down: the X-ray image enters the first station of the network, which looks for basic features such as edges or shapes. Then, it moves to subsequent stations that get more specific—maybe one station focuses on patterns related to lung texture, while another looks for signs of inflammation. Each station refines the information until the last one makes the call: healthy lungs or signs of pneumonia. The flow is like passing the X-ray through a series of checkpoints, each specializing in a different aspect relevant to the diagnosis.

An example could be classifying different types of tissue in MRI scans. The network starts by identifying general features in the scan and then progresses through layers to pinpoint specific details—such as distinguishing between muscle, bone, and organs. This "layered" analysis allows feedforward networks to efficiently process complex medical images, aiding in tasks such as tissue segmentation or disease identification with impressive accuracy (Fig. 4.2).

4.2.2 Convolutional Neural Networks (CNNs)

Convolutional neural networks (CNNs) are like super detectives specialized in understanding images. They break down pictures into smaller pieces and analyze them, much like how a detective examines various parts of a crime scene. In medical imaging, envision a CNN as an investigator inspecting MRI images to identify brain tumors (Fig. 4.3).

Here's how it works: when an MRI scan enters the CNN, it's broken down into tiny sections, where each section is analyzed to find patterns or features, such as textures or shapes, that are crucial for identifying abnormalities. For example, the CNN might "look" at various parts of the brain to spot irregular growths that could signal a tumor. By scrutinizing these smaller sections and piecing them together, the CNN figures out if the brain scan indicates a potential health issue.

A groundbreaking medical application could involve utilizing CNNs to analyze brain MRI scans for the early detection of neurodegenerative diseases such as Alzheimer's [5]. These neural networks can meticulously examine the intricate structures of the brain, pinpointing subtle alterations or irregularities that may signal the onset or advancement of neurocognitive disorders. This technology enables a targeted analysis of specific regions in the MRI where signs of atrophy or abnormal patterns may be indicative of neurodegenerative conditions.

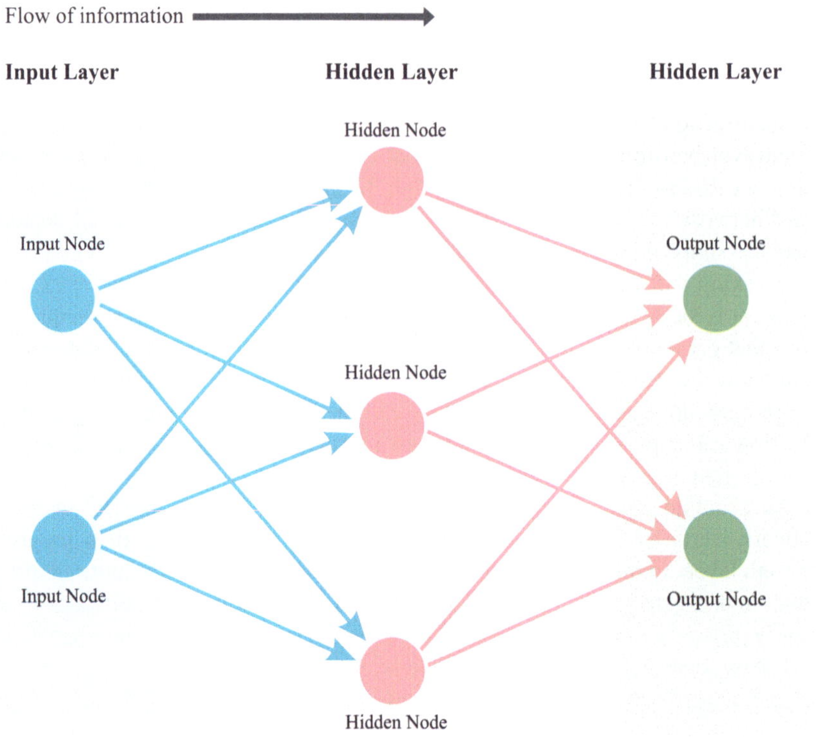

Fig. 4.2 Feedforward neural network. See text for further explanation

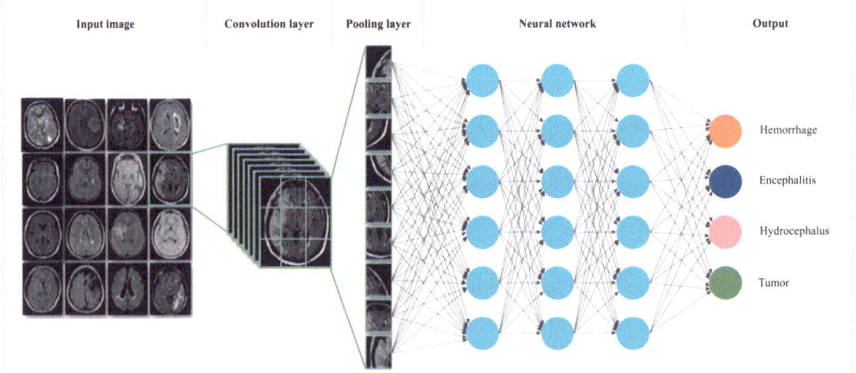

Fig. 4.3 Basic CNN architecture. See text for further explanation

4.2.3 Recurrent Neural Networks (RNNs)

Recurrent neural networks (RNNs) are a bit different from CNNs in that they're more like smart time travelers. While CNNs are great at understanding images by looking at different parts, RNNs are specialized in understanding sequences or data that come in a particular order, just like looking at a series of events over time [6]. In medicine, think of RNNs as experts analyzing a patient's heart rate data over time to predict potential heart issues.

Here's the key: RNNs have memory, so they remember previous bits of information they've seen. Imagine looking at a patient's heart rate history—a series of beats—and trying to spot irregularities. RNNs excel at this because they remember the patterns of normal heartbeats they've seen before. When a new beat comes in, they compare it to what they've learned, raising an alert if something seems off. This ability to remember and analyze sequences makes RNNs incredibly useful in tasks involving time-series data, such as predicting patient outcomes based on their medical history (Fig. 4.4).

An interesting medical application for RNNs could be in predicting epileptic seizures. By analyzing electroencephalogram (EEG) data, which shows brain activity over time, RNNs can learn patterns leading up to a seizure [7]. As they "watch" the sequence of brain waves, they can recognize the telltale signs that a seizure might occur, giving doctors a heads-up to intervene or prepare accordingly.

4.2.4 Autoencoders and Unsupervised Learning

Autoencoders are like clever artists that learn to capture the essence of things without needing labels or instructions. In simple terms, they're really good at learning how to represent and recreate data without being told what each piece means. In

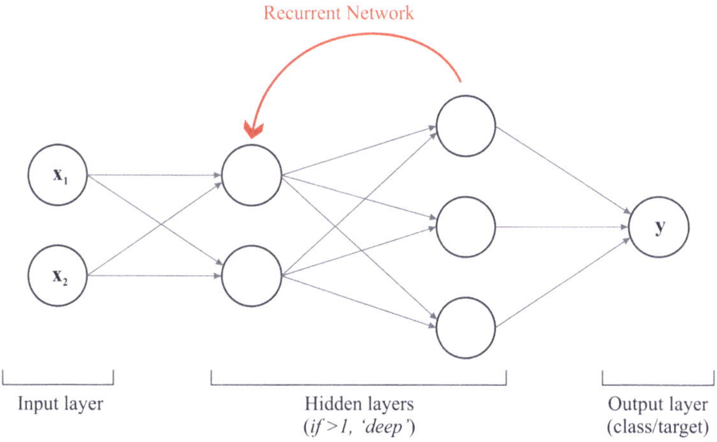

Fig. 4.4 Recurrent neural network (RNN). See text for further explanation

medical imaging, an autoencoder might be trained to reconstruct clear and noise-free images from messy X-rays or MRI scans [8].

Here's how it works: imagine handing an artist a blurry picture and asking them to make it clearer without knowing what the original looked like. Autoencoders do something similar—they're given messy or noisy medical images and learn to clean them up without being explicitly told what each detail means. They analyze the patterns in the data and figure out how to reconstruct a more precise and understandable version. For instance, in medical applications, an autoencoder might take an unclear ultrasound image and reconstruct it to highlight specific structures, helping doctors get a better view for diagnosis.

One unique medical application could involve using autoencoders for anomaly detection in medical images. For instance, they can learn to reconstruct "normal" X-ray images and notice deviations when presented with an image containing an anomaly, such as a fractured bone or an abnormal growth. This ability to identify what doesn't fit the learned "normal" patterns makes autoencoders valuable in unsupervised learning, aiding in anomaly detection and highlighting potential health issues within medical images.

4.2.5 Generative Adversarial Networks (GANs)

Generative adversarial networks (GANs) represent a revolutionary class of artificial intelligence algorithms introduced by Ian Goodfellow and his colleagues in 2014 [9]. GANs consist of two neural networks, a generator and a discriminator, engaged in a competitive learning process. The generator aims to produce realistic data, such as images, while the discriminator evaluates the generated samples against real ones [9, 10]. The adversarial dynamic between these networks drives continuous improvement, pushing the generator to create more convincing outputs and the discriminator to enhance its discrimination abilities.

The training process involves minimizing the generator's error in generating authentic-looking data and maximizing the discriminator's accuracy in distinguishing real from generated samples. GANs have demonstrated remarkable success in various domains, including image and video synthesis, style transfer, and even generating entirely new content. Despite their success, challenges such as mode collapse and training stability persist, making ongoing research essential for unlocking the full potential of GANs.

4.2.6 Attention Mechanisms and Transformers

Attention mechanisms and transformers function as smart assistants that concentrate on the most important parts of a task. In simple terms, envision reading a long document while a helpful guide highlights the crucial sentences or sections.

Similarly, attention mechanisms in transformers help neural networks focus on specific parts of data, emphasizing important details. In medicine, this technology can be crucial when analyzing complex medical images such as CT scans or X-rays.

Consider analyzing a chest X-ray to detect potential lung issues. Instead of the network looking at the entire image all at once, the attention mechanism in transformers guides it to focus on critical regions, such as specific areas showing signs of infection or abnormalities [11]. It's akin to a spotlight highlighting important sections of the X-ray—such as areas indicating tumors or lung cancer. By directing the network's attention, transformers enhance its ability to accurately identify and diagnose medical conditions from intricate imaging data.

A unique medical application for attention mechanisms in transformers could be seen in pathology slides. When examining tissue samples under a microscope, the network can use attention mechanisms to focus on regions that might indicate the presence of cancerous cells or anomalies, aiding pathologists in faster and more precise diagnoses. This ability to concentrate on specific areas within medical data amplifies the accuracy and efficiency of diagnostic processes, offering immense potential for improving healthcare outcomes.

In the domain of medical imaging, the comprehensive understanding and application of neural network fundamentals and their specialized iterations such as feedforward neural networks, CNNs, RNNs, autoencoders, GANs, and the advanced mechanisms of attention and transformers are vital for healthcare professionals. These tools serve as the backbone of cutting-edge diagnostic methodologies, allowing for nuanced analysis and interpretation of intricate medical images. From pinpointing anomalies in X-rays with CNNs to predicting disease progressions using RNNs and refining image clarity through autoencoders, each architecture plays a vital role. The integration of these deep learning techniques not only augments the precision and efficiency of diagnoses but also reshapes the landscape of medical care. It promises improved treatment planning, accurate disease identification, and ultimately better patient outcomes.

4.3 Convolutional Neural Networks (CNNs) for Image Classification

Convolutional neural networks (CNNs) stand as the backbone of modern image classification, revolutionizing how computers "see" and understand visual information. These specialized networks act as sleuths for images, enabling computers to spot patterns, shapes, and features within pictures, empowering them to distinguish between different objects or identify specific details within complex images [12]. In fields such as medicine, where precise identification of diseases from X-rays, MRI scans, or pathology images is critical, CNNs play a pivotal role by providing accurate and reliable analysis, enhancing diagnostic accuracy, and ultimately improving patient care (Fig. 4.5).

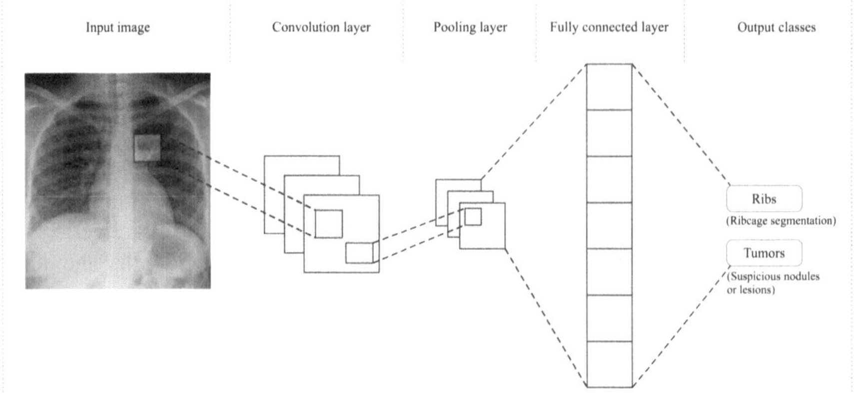

Fig. 4.5 CNN architecture for image classification. See text for further explanation

4.3.1 Motivation for CNNs in Image Classification

The motivation for CNNs in image classification lies in their inherent structure, which allows them to automatically learn and understand complex patterns in visual data without any human supervision [13]. CNNs are uniquely designed to process grid-like data, where the spatial relationships in the data are highly relevant. This makes them particularly effective for image classification tasks, where pixel values and their spatial locations in relation to other pixels are of central importance. This is achieved by using convolutional layers, which are the core building blocks of a CNN [14].

4.3.2 Convolutional Layers

Convolutional layers are responsible for extracting features from the input data, such as images. They consist of a collection of convolutional filters (also called kernels) that are applied to the input data with each filter designed to detect a specific type of feature in the image, such as a line of a specific orientation and a blob of a certain size. This results in the generation of output feature maps. These feature maps summarize the presence of detected features in the input, such as edges, colors, and other visual elements [15]. Thus, convolutional layers are typically responsible for producing a set of feature maps that highlight the locations in the image where these features were detected.

4.3.3 Pooling Layers for Downsampling

Pooling layers are used to downsample the feature maps generated by the convolutional layers. They provide an approach to summarize the presence of features in patches of the feature map, making the resulting downsampled feature maps more robust to changes in the position of the feature in the image. This capability added by pooling is called the model's invariance to local translation [16].

These pooling layers aim to reduce the spatial dimensions of the input data (width and height) while preserving the most important feature information. This helps to decrease the computational cost of training the CNN and controls overfitting by providing a form of translation invariance. Common pooling methods include max pooling, which takes the maximum value in a certain region, and average pooling, which takes the average value.

For instance, in an aerial view photo of a landscape, one may encounter the task of identifying distinct features, such as buildings, roads, and trees. The image, being large and detailed, poses a challenge in focusing on specific elements.

Pooling layers in a neural network can help by downsampling the image. It's like looking at the photo through a smaller window, reducing the overall size but retaining the essential information. For instance, using a pooling technique such as max pooling, the network takes smaller sections of the image (e.g., a 2 × 2 grid of pixels) and selects the maximum value within each section. This downsizes the image by combining and preserving the most significant details.

In this analogy, after applying pooling, the image becomes smaller while maintaining crucial features such as major roads, building shapes, and large trees. This downsampling helps the neural network focus on important aspects of the landscape, aiding in quicker and more efficient analysis for tasks such as identifying landmarks or significant elements within the photo.

4.3.4 Filter Sizes and Depth

Filter sizes refer to the dimensions of the convolutional filters used in the convolutional layers. These filters are applied to the input data to create feature maps, and the size of the filter determines the receptive field of the neuron [17]. The depth of a CNN refers to the number of layers in the network, with each layer focusing on different levels of features.

To compare filter sizes and depth in a CNN, consider examining different magnifying glasses when analyzing an image.

Filter sizes refer to the dimensions of the "magnifying glass" the CNN uses to analyze parts of an image. For example, a larger filter size, such as 5 × 5 pixels, covers a more extensive area of the image at once. Conversely, a smaller filter size, such as 3 × 3 pixels, scrutinizes a smaller portion. This impacts what features the network can detect. Consider trying to find details in a picture—using a larger magnifying glass might assist in spotting broad shapes, while a smaller one allows focusing on finer details.

Moreover, the depth in a CNN relates to how many different "magnifying glasses" or filters the network uses. Having more depth means employing numerous types of magnifying glasses of various sizes (e.g., 3 × 3, 5 × 5, or even larger). This depth allows the network to detect a wide range of features, from larger structures such as shapes or edges (using larger filters) to finer details such as textures or complicated patterns (using smaller filters). It's like having a set of magnifying glasses with different strengths to inspect various aspects of the image, helping the network understand both general and intricate features for accurate analysis in tasks such as recognizing objects or abnormalities in medical images.

In essence, filter sizes and depth in the convolutional layers significantly affect the performance of CNNs. The filter size determines the size of the features that the layer can detect, while the depth (number of filters) determines the variety of features that can be detected. A deep CNN can model complex patterns by combining simpler features from the lower layers into more abstract features in the higher layers.

4.3.5 Activation Maps and Receptive Fields

Activation maps and receptive fields are integral parts of understanding how a CNN processes an image. An activation map is the output of each convolutional or pooling layer, showing which features were detected and where. The receptive field of a neuron in a layer is the region of the input image that contributes to the neuron's activation [18]. As one moves deeper into the network, the receptive field of a neuron increases, allowing it to respond to larger regions of the input image. In the context of CNNs, the receptive field of a neuron is determined by the convolutional filter it uses.

To examine a chest X-ray, it is important to explore how activation maps and receptive fields in a neural network can be applied in medical imaging.

Activation maps could highlight regions in the chest X-ray where specific features related to certain lung conditions or abnormalities are detected. For example, if the network is configured to recognize signs of chronic obstructive pulmonary disease (COPD), the activation map might spotlight areas showing emphysematous changes or bronchial wall thickening. In the case of lung cancer detection, the map could emphasize nodules or masses indicative of malignancy. For interstitial lung diseases such as pulmonary fibrosis, the activation map may highlight patterns of fibrotic tissue or honeycombing. These highlighted regions help clinicians focus on potential areas of concern, aiding in the diagnosis process by drawing attention to potential abnormalities.

Now, receptive fields in this context would define the area of the X-ray that each "observer" in the neural network "sees" or pays attention to. A smaller receptive field might focus on a specific portion of the lung, inspecting minute details such as subtle changes in tissue density or texture associated with pulmonary fibrosis. In contrast, a larger receptive field might encompass a more comprehensive view of the lung, capturing broader patterns or structures relevant to the diagnosis, such as the presence of consolidations or infiltrates in different lung regions.

So, in the realm of medical imaging, activation maps and receptive fields in a neural network help highlight potential abnormalities or disease-related features within images such as X-rays. This assists healthcare professionals in identifying specific regions of interest for accurate diagnosis and treatment planning.

4.3.6 Transition to Fully Connected Layers

Fully connected layers (FC layers) are the final layers of a CNN, responsible for making high-level abstractions of the input features. They receive the mid- and low-level features from the previous layers and create the high-level abstraction, which represents the last-stage layers of a typical neural network. The classification scores are generated using the ending layer, with each score representing the probability of a specific class [19].

Now relate the transition to fully connected layers in a neural network to interpreting data from medical images, such as a brain MRI scan.

In a brain MRI analysis using a CNN, after the network identifies various patterns, textures, or structures within the image (such as detecting specific regions or abnormalities), it reaches the phase of transitioning to fully connected layers. This transition involves gathering and combining information from these identified features to make a final decision or diagnosis.

For instance, consider that the CNN has identified multiple areas in the brain MRI scan that might indicate different conditions or anomalies, such as tumorlike structures or abnormal tissue patterns. The fully connected layers in the network gather this information from these different regions and features, integrating and processing these insights to make a conclusive assessment about the presence, location, or characteristics of potential abnormalities in the brain. This final integration of information from various parts of the image assists healthcare professionals in making informed decisions about the patient's condition. It aids in accurate diagnoses and treatment planning based on the comprehensive analysis provided by the network.

Thus, transition to fully connected layers is the final step in a typical CNN architecture. After several iterations of convolutional and pooling layers, the final feature maps are flattened and fed into one or more fully connected layers, which perform high-level reasoning on the features. These layers use the processed information from previous layers to classify the image (Fig. 4.6).

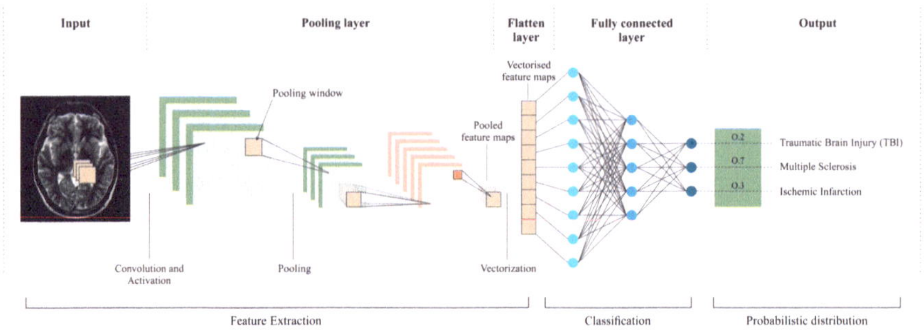

Fig. 4.6 Full CNN representation. See text for further explanation

4.3.7 Dropout and Regularization

In deep learning models, especially neural networks, overfitting is a common concern. Overfitting occurs when a model learns the training data too well, including its noise and outliers, leading to poor performance on new, unseen data. *Dropout* is a regularization technique designed to address this issue.

Imagine you have a neural network with multiple layers, each containing numerous neurons. During training, the dropout technique involves randomly deactivating (or "dropping out") a certain percentage of neurons at each iteration. This dropout is applied independently to each neuron, meaning that different neurons are dropped out in each training iteration [20].

For instance, let's consider a simple neural network with three layers: an input layer, a hidden layer, and an output layer. The hidden layer has multiple neurons, and we decide to apply a dropout rate of 20%. This means that, during each training iteration, 20% of the neurons in the hidden layer will be randomly deactivated. The network might look something like this:

- Input layer.
- Hidden layer (with 20% of neurons dropped out).
- Output layer.

By randomly dropping out neurons, the network is forced to adapt and learn different redundant representations of the data. This helps prevent the reliance on specific neurons and features, making the model more robust and less likely to overfit the training data.

The process of dropout encourages the remaining neurons to work more independently and not rely too heavily on specific features, which enhances the model's ability to generalize well to unseen data. During the testing or inference phase, all neurons are usually active, as dropout is only applied during training.

Regularization is another technique used to prevent overfitting by adding a penalty term to the loss function, which encourages the model to use simpler features and avoid overly complex ones. Regularization techniques, such as L1 and L2, add a penalty to the loss function based on the size of the network's weights, encouraging the model to keep the weights small and reduce overfitting [21].

For instance, imagine the CNN is learning to identify signs of tuberculosis in chest X-rays. Dropout involves randomly "switching off" or ignoring a certain percentage of neurons or connections in the network during training. It forces the network to learn more robust and generalized patterns across the image, ensuring that it doesn't overly focus on individual aspects that might not consistently indicate the presence of the condition.

Regularization, on the other hand, penalizes overly complex or extreme patterns learned by the network during training. It prevents the network from becoming too sensitive to small variations or noise in the X-ray images that might not be clinically relevant. For example, regularization could help in reducing the network's sensitivity to irrelevant shadows or artifacts that might appear in the X-rays but don't necessarily signify the presence of tuberculosis. By encouraging the network to focus on essential features and avoid extreme patterns, dropout and regularization techniques enhance the network's ability to generalize and make more accurate diagnoses in medical imaging tasks.

4.3.8 Transfer Learning with Pretrained CNNs

Transfer learning is a technique where pretrained CNNs are used as a starting point for new tasks. This approach leverages the knowledge learned by the CNN from a large dataset, allowing for faster and more accurate model training on new tasks. Fine-tuning and optimizing a pretrained model and leveraging transfer learning emerge as widely embraced methodologies in deep learning, particularly valuable when there's a scarcity of labeled data [22]. This significantly reduces the amount of labeled data required and often results in superior performance due to the rich feature representations learned by the pretrained model.

Consider the application of transfer learning with pretrained CNNs in medical imaging for the detection of abnormalities in retinal images. In this scenario, a CNN has been pretrained on a diverse dataset containing images of various objects, scenes, and patterns. When faced with the task of identifying retinal abnormalities, transfer learning becomes a valuable technique.

Imagine the pretrained CNN has already learned fundamental visual features, such as edges, shapes, and textures that are common across different types of images. Instead of training a new CNN from scratch for retinal image analysis, which would require a substantial amount of specific medical imaging data, the knowledge gained from the pretrained CNN can be leveraged.

By fine-tuning the pretrained CNN on a smaller dataset of retinal images, the network can adapt its learned features to detect abnormalities such as macular degeneration. The model can use its prior knowledge to identify patterns indicative of retinal diseases, aiding ophthalmologists in making early and accurate diagnoses. This approach accelerates the model training process and enhances the accuracy of the system for retinal image analysis by building on the foundational knowledge embedded in the pretrained CNN.

4.3.9 Data Augmentation Strategies

Data augmentation is a technique used to increase the variety of training data, which helps improve the generalization of the model and reduces overfitting. It involves creating new training examples by applying transformations such as rotation, scaling, flipping, and elastic deformation to the original images [23]. Data augmentation significantly enhances model performance, particularly when there's a limited availability of labeled data.

Consider data augmentation strategies in the context of classifying different types of cells in a microscope image to aid in disease diagnosis.

In a medical setting, especially in pathology, microscopic images of cells play a crucial role. Data augmentation involves manipulating and expanding the available dataset by applying various transformations to these images without altering their diagnostic features.

For instance, imagine a dataset of microscopic images of blood cells used to train a CNN for cell classification. Data augmentation strategies might involve rotating the images slightly, flipping them horizontally or vertically, changing the brightness or contrast, or applying slight shifts or distortions to simulate different viewing angles or conditions.

These transformations create new versions of the original images, increasing the diversity of the dataset. This augmented dataset helps improve the CNN's ability to generalize and recognize various cell types, contributing to more accurate disease diagnoses based on microscopic cell analysis.

4.3.10 Interpretability of CNNs

The interpretability of CNNs pertains to comprehending the model's decision-making process and elucidating its outputs in a manner understandable to humans. Methods such as visualizing activation maps, assessing the significance of individual features, and employing explanation algorithms serve to interpret the decision-making of a CNN [24]. Enhancing interpretability involves techniques such as visualizing feature maps, revealing the learned features of the networks, and utilizing approaches such as occlusion and saliency maps to highlight crucial portions of the image for the model's prediction [25, 26]. Interpretability remains a crucial element in deep learning, aiding users in grasping their models' behavior and enhancing overall performance.

Imagine using a CNN for medical image classification, such as identifying whether an X-ray depicts a healthy or fractured bone.

The interpretability of CNNs refers to understanding why and how the network makes its predictions. Just like a doctor explains why they diagnose a fracture based on certain X-ray features, interpretability in CNNs involves understanding which parts of the X-ray image are crucial for the network to make its decision.

For instance, if the CNN identifies a fracture, interpretability could help highlight the regions or features (such as specific bone structures or fractures) that the network focused on to arrive at that conclusion. This insight can be crucial for doctors to trust the AI's diagnosis, understand its reasoning, and potentially catch nuances that the AI might have missed or misinterpreted.

Methods for interpretability might involve visualizations showing the areas of the X-ray image that the CNN deemed most important. It helps doctors corroborate the AI's findings and gain confidence in using AI as an aid in diagnosis.

4.4 Architectures for Segmentation, Object Detection, and Image Generation

In the realm of medical imaging, the development of sophisticated architectures for segmentation, object detection, and image generation has emerged as a pivotal frontier, advancing diagnostic and analytical capabilities. As technology continues to advance, the demand for precise and efficient tools in medical image processing intensifies, prompting the exploration of innovative methodologies.

This section encapsulates the intricate world of designing frameworks that facilitate the extraction of meaningful information from medical images. It enables accurate segmentation of structures, robust detection of objects of interest, and the generation of high-quality images. The convergence of deep learning, computer vision, and medical imaging has given rise to these architectures that have had transformative impact on healthcare practices.

4.4.1 Segmentation Architectures

Segmentation in medical imaging involves extracting regions of interest (ROIs) from 3D image data, such as from MRI or CT scans. This process allows for a more precise analysis of anatomical data by isolating only necessary areas, such as simulating physical properties or virtually positioning CAD-designed implants within a patient [8, 27]. U-Net and fully convolutional networks (FCNs) are popular architectures for segmentation in medical imaging.

The following section delves deeper into each of them to gain a more comprehensive understanding.

4.4.1.1 U-Net Architecture

U-Net is a convolutional network architecture designed for fast and precise segmentation of images. It has been widely used in medical image segmentation and has shown advantages in accurately segmenting small targets. U-Net utilizes a U-shaped structure with skip connections to combine high-level features with low-level features, enabling precise segmentation of medical images [28]. It has outperformed prior methods in challenges for segmentation of neuronal structures in electron microscopic stacks. It has also been used for tasks such as 3D printing patient-specific models from MRI scans [29–31].

In the U-Net architecture, skip connections are used to address the challenge of information loss during the downsampling (contracting) and upsampling (expanding) paths of the network. The network consists of a contracting path (encoder) and an expanding path (decoder). The contracting path involves the repeated application of convolutional layers and pooling operations, which reduces the spatial resolution and captures high-level features. The expanding path involves upsampling the features back to the original resolution while incorporating contextual information.

Skip connections are introduced by connecting the layers at the same resolution between the contracting and expanding paths. Before each downsampling step (usually a pooling operation), the current feature map is saved or concatenated with the feature map at the same resolution in the contracting path. This saved information is then used in the expanding path to recover spatial details lost during downsampling.

Thus, the skip connections enable the network to maintain both low-level and high-level features throughout the architecture, thereby preserving the fine-grained details that allow the U-Net to achieve accurate segmentation results. Moreover, the architecture resembles a "U" shape, hence the name U-Net.

U-Net, devised by Olaf Ronneberger, Philipp Fischer, and Thomas Brox in 2015, was initially designed to excel in biomedical image segmentation but has since proven its adaptability and effectiveness across various disciplines such as autonomous vehicles and computer vision, establishing itself as an immensely powerful asset [32].

Tailored primarily for processing image data, U-Net demonstrates its flexibility by accommodating an array of image inputs, spanning from grayscale to color images. Within the realm of biomedical applications, U-Net stands out, particularly skilled at handling intricate microscopy images, CT scans, and MRI images. Moreover, its adaptability extends to handling multichannel images or volumetric 3D data, showcasing its versatility in diverse imaging scenarios.

In assembling a jigsaw puzzle where each piece represents a part of an organ in a medical image, the U-Net functions as a master puzzle solver. It is designed to comprehend not only the edges or corners of the pieces but also the intricate details within them. For example, in brain imaging, when segmenting different parts such as the white matter, gray matter, and various structures, the U-Net excels at precisely outlining these different regions with high accuracy (see Fig. 4.7). Its design allows it to capture fine details while maintaining a broader understanding of the overall context.

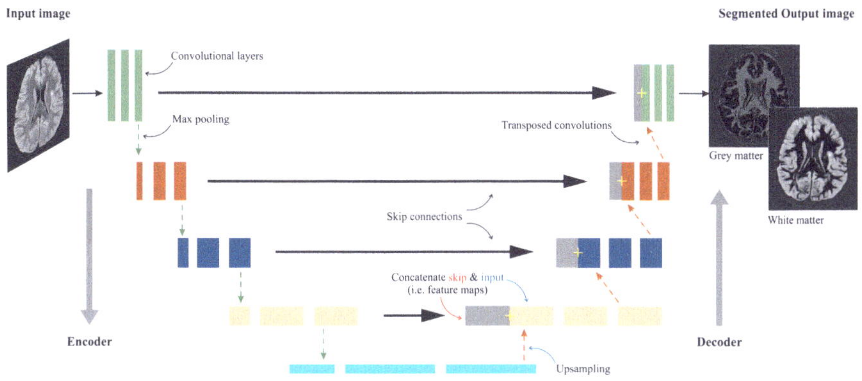

Fig. 4.7 U-Net architecture for segmenting brain parts (gray matter, white matter). See text for further explanation

4.4.1.2 Fully Convolutional Networks (FCNs)

Fully convolutional networks (FCNs) are a class of neural networks designed for pixel-level tasks such as semantic segmentation, enabling the detailed understanding of visual data. Unlike traditional convolutional neural networks that incorporate fully connected layers for classification, FCNs preserve spatial information by replacing these fully connected layers with convolutional ones. This architectural adaptation allows FCNs to process images of varying sizes and produce outputs with the same spatial dimensions as the input [33].

In practical terms, consider an FCN applied to an image segmentation task, such as identifying different objects within a scene. The FCN's architecture involves multiple layers of convolutions that extract features at different levels of abstraction from the input image. The network then performs upsampling using transposed convolutions or other techniques to regain the original spatial resolution. This generates a dense output where each pixel corresponds to a particular class or object label. This approach allows FCNs to accurately segment and delineate objects in an image, making them particularly valuable in applications such as object detection, medical image analysis, and scene understanding (Fig. 4.8).

In the depicted network architecture, the initial input image with dimensions H x W undergoes a series of convolutional operations, gradually reducing its spatial dimensions to $H/2 \times W/2$ and eventually to $H/4 \times W/4$. At this point, the network generates smaller-scale heatmaps representing various objects within the image, where the intensity of each pixel reflects the likelihood of the presence of a particular object. Subsequently, these smaller heatmaps undergo an upsampling process and are then combined to produce a final high-resolution segmentation map. In this map, every pixel is assigned to the class with the highest probability, effectively delineating and categorizing objects across the entire original image with increased detail and precision.

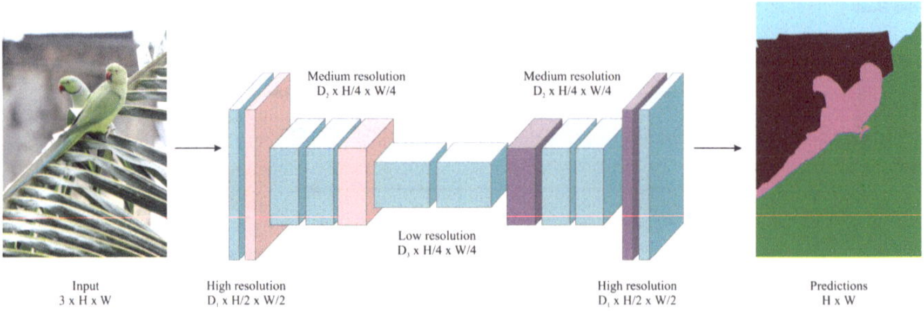

Fig. 4.8 Fully convolutional network (FCN) architecture. See text for further explanation

Picture trying to color in a drawing using different color codes for various objects. FCNs operate like magic color-filling pens that discern the appropriate color for each segment of the image without requiring explicit instructions. In medical imaging, FCNs excel at pixel-level understanding. For instance, in a chest X-ray, FCNs can identify and segment areas with abnormalities such as bronchitis or lung nodules by labeling each pixel as part of the affected region or not. Their strength lies in their ability to see and classify every pixel in an image, which helps in detailed and accurate segmentation.

Both U-Net and FCNs are powerful tools used for image segmentation. U-Net excels at capturing detailed structures and context, while FCNs specialize in pixel-level understanding. They both aim to precisely outline specific areas or structures within medical images, assisting medical professionals in accurate diagnosis and treatment planning by highlighting crucial details that might otherwise be challenging to discern.

Along similar lines, the full resolution convolutional neural network (FrCN) is another important segmentation architecture employed in medical imaging [34]. FrCN is a type of deep learning model that aims to process and analyze medical images, such as MRI scans, CT scans, or X-rays, at their original resolution without downsampling or losing details.

In medical imaging, the use of FrCN is particularly important because it allows for the retention of fine structural details that can be critical for accurate diagnosis and treatment. By processing images at full resolution, FrCN can potentially improve the accuracy of detection, classification, and segmentation of medical conditions, leading to more reliable and precise medical assessments.

FrCN models are designed to handle the complexity and high-dimensional nature of medical images, enabling healthcare professionals to leverage advanced computational techniques in image analysis and interpretation. These models can facilitate tasks such as tumor detection, organ segmentation, disease classification, and various other diagnostic and predictive analyses.

Overall, the use of FrCN contributes to the advancement of computer-aided diagnosis and the overall quality of healthcare by enhancing the capabilities of image analysis and extraction of clinically relevant information from medical images.

4.4.2 *Object Detection Architectures*

Object detection in medical images aims to identify the location and classify different objects within the images. This is crucial for tasks such as lesion location, tracking, and image discrimination. Popular architectures for object detection in medical imaging include region proposal network (RPN), single-stage models such as You Only Look Once (YOLO), and two-stage models such as faster R-CNN and mask R-CNN [35]. A more recent approach is the MS transformer (multiscale), which uses a self-supervised learning approach to perform random masking on the input image and introduces a hierarchical transformer model for better object detection performance [36].

The following section dives into the core principles behind RPNs and single-shot detectors (SSDs) in medical imaging using straightforward examples:

4.4.2.1 Region Proposal Networks (RPNs)

Region proposal networks (RPNs) play a crucial role in extracting multiscale features and suggesting regions of interest within an image. These networks are frequently employed in two-stage detection models such as faster R-CNN and mask R-CNN. In the case of mask R-CNN, an additional module is integrated to simultaneously predict the segmentation mask, bounding box regression, and classification for the identified region of interest [37–39]. RPNs utilize a fully convolutional network (FCN) for the segmentation module, connecting neurons from one layer to the next for effective feature representation.

Think of RPNs as efficient scouts in a treasure hunt. Imagine a map representing a medical image with various potential areas of interest, such as tumors or anomalies. RPNs are much like skilled scouts who quickly identify and mark spots on the map where treasure might be hidden. In medical imaging, RPNs do something similar by proposing regions in an image where objects or abnormalities might exist. For instance, in mammograms, RPNs might swiftly suggest regions where potential tumors could be located, narrowing down the search area for further analysis. Their primary role is to propose these candidate regions efficiently and accurately.

Consider an example of faster R-CNN used in object detection, as explained in the March 2018 paper titled "Detecting and classifying lesions in mammograms with Deep Learning" by Ribli et al. [40] In this approach, there's a special set of convolutional layers, specifically the "region proposal network (RPN)." These layers are added on top of the last convolutional layer of the original network. Their job is to find and pinpoint objects in an image, regardless of the object's type.

To achieve this, the system uses default detection boxes of various sizes and shapes. These boxes help identify objects with different dimensions. The region proposals, or the most promising boxes, are then forwarded to the next stage of the network.

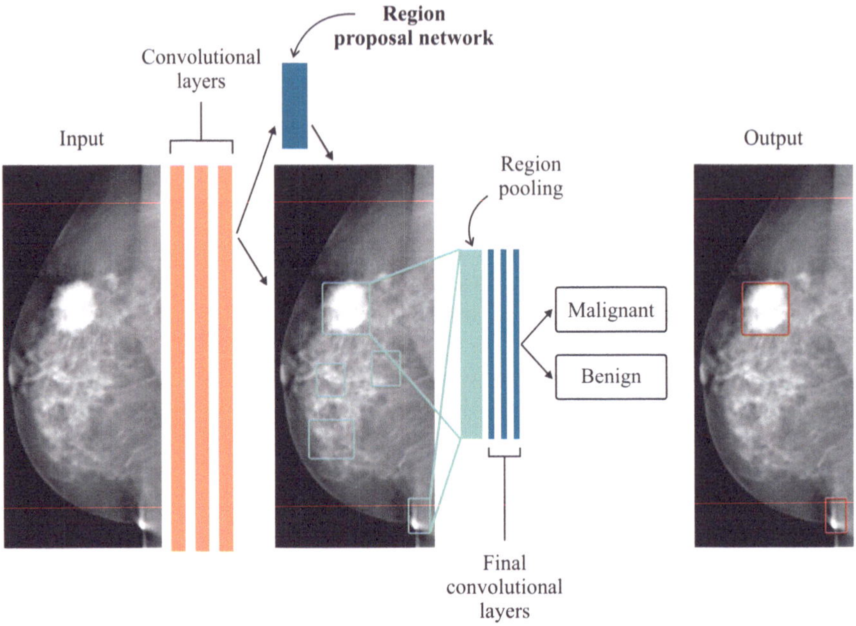

Fig. 4.9 Faster R-CNN showing region proposal network [40]. See text for further explanation

In the next part of the network, each proposed region is carefully analyzed. The region is resized to a fixed dimension, and the network tries to figure out if there's an object present and what class it belongs to. Additionally, it refines the boundaries of the detected object.

The model aims to classify objects and fine-tune their boundaries. It identifies objects such as benign or malignant lesions in medical images. The model outputs a bounding box for each detected lesion and a score, indicating how confident the model is in its classification. The final step involves eliminating redundant predictions using a process called non-maximum suppression. The overall model structure is illustrated in Fig. 4.9.

In summary, this approach helps the system understand and classify objects in images, such as lesions in mammograms, providing valuable information about their location, type, and confidence level.

4.4.2.2 Single-Shot Detectors (SSDs)

On the other hand, single-shot detectors (SSDs) such as YOLO use a fully convolutional approach to find all objects within an image in one pass through the network [41]. SSDs are more efficient and have good accuracy, making them suitable for medical imaging applications. For example, in a study focused on detecting brain metastases, deep learning SSDs were utilized to map axial MRI slices and generate bounding box predictions that encapsulate metastases along with associated detection confidence scores [42].

Consider organized items placed in different compartments of a toolbox. SSDs are like a quick, comprehensive glance at the toolbox, instantly identifying what's in each compartment without needing to open them individually. Similarly, in medical imaging, SSDs excel at detecting multiple objects or abnormalities in a single pass through the image. For instance, in a CT scan, SSDs can rapidly identify and categorize various anomalies such as fractures or nodules without needing multiple rounds of analysis. They efficiently scan the entire image, detecting different objects or abnormalities in one go.

Thus, both RPNs and SSDs play a crucial role in object detection, enabling the identification and classification of objects within images. These architectures are often used in combination with other deep learning models to improve the overall performance of object detection in medical imaging.

4.4.3 Image Generation Architectures

Image generation in medical imaging involves creating synthetic images from real or hybrid data. This process can help in better understanding of medical conditions. Popular architectures for image generation include generative adversarial networks (GANs), deep convolutional generative networks (DCGANs), and variational autoencoders (VAEs) [43, 44]. These networks can be used for various tasks, such as MRI reconstruction from given reference images, low-dose CT denoising, and generating synthetic eye fundus images [45].

The detailed exploration of some fundamental architectures used for generating images are discussed in the following section.

4.4.3.1 Generative Adversarial Networks (GANs)

As previously mentioned, GANs have frequently been employed to create realistic medical images, tackling the challenge of limited availability of diverse and authentic datasets crucial for comprehensive medical imaging research and studies. This has been particularly beneficial for behavioral experiments on medical image perception, where genuine and easily controllable medical images are critical. GANs have been applied to various image modalities, including mammograms, MRI, CT, and skin cancer images, with successful human evaluations verifying their effectiveness [46] (Fig. 4.10)

Imagine a scenario where an artist is trying to create lifelike paintings, and an art critic is trying to distinguish between real masterpieces and fakes. The artist (the generator) keeps producing paintings, aiming to fool the critic (the discriminator) into thinking they're real. As this creative contest goes on, the artist improves their

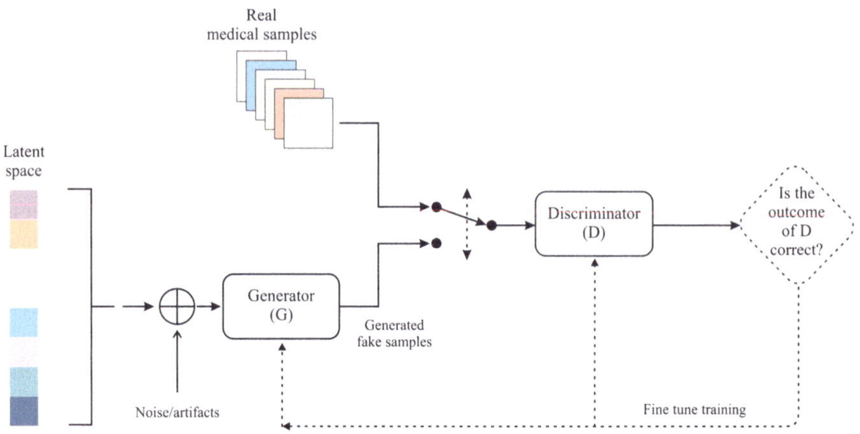

Fig. 4.10 Generative adversarial networks (GANs). See text for further explanation

skills to make paintings that look more and more authentic, while the critic becomes better at spotting the fakes. In the end, the artist generates paintings that are incredibly realistic, almost indistinguishable from the real ones.

In medical imaging, GANs work similarly. There are two components: a generator and a discriminator. The generator creates synthetic images resembling real medical data, such as MRI scans or X-rays. Meanwhile, the discriminator examines these generated images, trying to differentiate between real and fake ones. Through this back-and-forth process, the generator improves its ability to create more realistic medical images, enhancing its accuracy over time. Thus, GANs could generate synthetic MRI scans resembling real patient data, aiding in augmenting datasets for training AI models without needing extensive real data.

4.4.3.2 Variational Autoencoders (VAEs)

Variational autoencoders (VAEs) are a type of neural network known for their versatility and application in various tasks, particularly in the field of medical imaging. These networks are rooted in probability theory and have demonstrated effectiveness in tasks such as image segmentation, classification, and registration within the medical imaging domain [47].

In essence, VAEs function like artists exploring and capturing the essence of a particular painting style. Think of an artist who creates a unique style by blending various artistic techniques. Similarly, in medical imaging, VAEs aim to capture the essential features and variations present in a dataset. They achieve this by learning a lower-dimensional representation of input images, reducing noise, and enhancing the accuracy of subsequent tasks.

To illustrate, consider the analogy of an artist creating variations of a painting. In medical imaging, VAEs strive to understand and represent the underlying distribution of data points in a dataset. This knowledge enables them to generate new, realistic images with slight variations. For instance, when tasked with reconstructing missing parts of an image or generating diverse versions of a specific medical scan, VAEs can produce outputs that vary while maintaining the essential characteristics of the original medical images.

The practical applications of VAEs in medical imaging include their ability to generate new, plausible variations of existing images. This capability is valuable for tasks such as analysis and training, where having diverse and realistic datasets is crucial. Whether it involves reconstructing missing information or creating variations for training purposes, VAEs contribute significantly to enhancing the robustness and effectiveness of medical imaging processes.

Both GANs and VAEs are image generation architectures, but they work differently. GANs engage in a competitive game between a generator and a discriminator to create realistic images, while VAEs focus on learning and generating new data points within a learned distribution. Together, they enable the creation of synthetic medical images for various purposes, such as data augmentation, filling in missing information, or generating variations for training AI models. Ultimately, the process aides in more robust medical imaging analyses and advancements.

References

1. Sarker, I.H. Deep Learning: A Comprehensive Overview on Techniques, Taxonomy, Applications and Research Directions. SN COMPUT. SCI. 2, 420 (2021). https://doi.org/10.1007/s42979-021-00815-1
2. Montesinos López, O.A., Montesinos López, A., Crossa, J. (2022). Fundamentals of Artificial Neural Networks and Deep Learning. In: Multivariate Statistical Machine Learning Methods for Genomic Prediction. Springer, Cham. https://doi.org/10.1007/978-3-030-89010-0_10
3. Yalçın, O.G. (2021). Feedforward Neural Networks. In: Applied Neural Networks with TensorFlow 2. Apress, Berkeley, CA. https://doi.org/10.1007/978-1-4842-6513-0_6
4. Yagawa, G., Oishi, A. (2021). Feedforward Neural Networks. In: Computational Mechanics with Neural Networks. Lecture Notes on Numerical Methods in Engineering and Sciences. Springer, Cham. https://doi.org/10.1007/978-3-030-66111-3_2
5. Patil, V., Madgi, M. & Kiran, A. Early prediction of Alzheimer's disease using convolutional neural network: a review. Egypt J Neurol Psychiatry Neurosurg 58, 130 (2022). https://doi.org/10.1186/s41983-022-00571-w
6. Das, S., Tariq, A., Santos, T., Kantareddy, S.S., Banerjee, I. (2023). Recurrent Neural Networks (RNNs): Architectures, Training Tricks, and Introduction to Influential Research. In: Colliot, O. (eds) Machine Learning for Brain Disorders. Neuromethods, vol 197. Humana, New York, NY. https://doi.org/10.1007/978-1-0716-3195-9_4
7. Carrión, S., López-Chilet, Á., Martínez-Bernia, J., Coll-Alonso, J., Chorro-Juan, D., Gómez, J.A. (2022). Automatic Detection of Epileptic Seizures with Recurrent and Convolutional Neural Networks. In: Mazzeo, P.L., Frontoni, E., Sclaroff, S., Distante, C. (eds) Image Analysis and Processing. ICIAP 2022 Workshops. ICIAP 2022. Lecture Notes in Computer Science, vol 13373. Springer, Cham. https://doi.org/10.1007/978-3-031-13321-3_46
8. Iqbal, S., A. N. Qureshi, Li, J. et al. On the Analyses of Medical Images Using Traditional Machine Learning Techniques and Convolutional Neural Networks. Arch Computat Methods Eng 30, 3173–3233 (2023). https://doi.org/10.1007/s11831-023-09899-9

9. Ian J. Goodfellow, Jean Pouget-Abadie, Mehdi Mirza, Bing Xu, David Warde-Farley, Sherjil Ozair, Aaron Courville, Yoshua Bengio, Generative Adversarial Networks, June 2014, arXiv:1406.2661, https://doi.org/10.48550/arXiv.1406.2661
10. Rather, I.H., Kumar, S. Generative adversarial network based synthetic data training model for lightweight convolutional neural networks. Multimed Tools Appl 83, 6249–6271 (2024). https://doi.org/10.1007/s11042-023-15747-6
11. Yutong Xie, et al., Attention Mechanisms in Medical Image Segmentation: A Survey, 2023, URL: https://arxiv.org/pdf/2305.17937.pdf
12. Benali Amjoud A, Amrouch M. Convolutional Neural Networks Backbones for Object Detection. Image and Signal Processing. 2020 Jun 5;12119:282–9. https://doi.org/10.1007/978-3-030-51935-3_30. PMCID: PMC7340949.
13. Alzubaidi, L., Zhang, J., Humaidi, A.J. et al. Review of deep learning: concepts, CNN architectures, challenges, applications, future directions. J Big Data 8, 53 (2021). https://doi.org/10.1186/s40537-021-00444-8
14. IBM, What are convolutional neural networks? URL: https://www.ibm.com/topics/convolutional-neural-networks
15. Yamashita, R., Nishio, M., Do, R.K.G. et al. Convolutional neural networks: an overview and application in radiology. Insights Imaging 9, 611–629 (2018). https://doi.org/10.1007/s13244-018-0639-9
16. Nirthika, R., Manivannan, S., Ramanan, A. et al. Pooling in convolutional neural networks for medical image analysis: a survey and an empirical study. Neural Comput & Applic 34, 5321–5347 (2022). https://doi.org/10.1007/s00521-022-06953-8
17. Montesinos López, O.A., Montesinos López, A., Crossa, J. (2022). Convolutional Neural Networks. In: Multivariate Statistical Machine Learning Methods for Genomic Prediction. Springer, Cham. https://doi.org/10.1007/978-3-030-89010-0_13
18. Vakalopoulou, M., Christodoulidis, S., Burgos, N., Colliot, O., Lepetit, V. (2023). Deep Learning: Basics and Convolutional Neural Networks (CNNs). In: Colliot, O. (eds) Machine Learning for Brain Disorders. Neuromethods, vol 197. Humana, New York, NY. https://doi.org/10.1007/978-1-0716-3195-9_3
19. Alzubaidi L, Zhang J, Humaidi AJ, Al-Dujaili A, Duan Y, Al-Shamma O, Santamaría J, Fadhel MA, Al-Amidie M, Farhan L. Review of deep learning: concepts, CNN architectures, challenges, applications, future directions. J Big Data. 2021;8(1):53. https://doi.org/10.1186/s40537-021-00444-8. Epub 2021 Mar 31. PMID: 33816053; PMCID: PMC8010506.
20. Huawei Technologies Co., Ltd. (2023). Overview of Deep Learning. In: Artificial Intelligence Technology. Springer, Singapore. https://doi.org/10.1007/978-981-19-2879-6_3
21. IBM, What is regularization? URL: https://www.ibm.com/topics/regularization
22. Hosna, A., Merry, E., Gyalmo, J. et al. Transfer learning: a friendly introduction. J Big Data 9, 102 (2022). https://doi.org/10.1186/s40537-022-00652-w
23. Goceri, E. Medical image data augmentation: techniques, comparisons and interpretations. Artif Intell Rev 56, 12561–12605 (2023). https://doi.org/10.1007/s10462-023-10453-z
24. Teng, Q., Liu, Z., Song, Y. et al. A survey on the interpretability of deep learning in medical diagnosis. Multimedia Systems 28, 2335–2355 (2022). https://doi.org/10.1007/s00530-022-00960-4
25. Elhassan Mohamed, Konstantinos Sirlantzis, Gareth Howells, A review of visualisation-as-explanation techniques for convolutional neural networks and their evaluation, Displays, Volume 73, 2022, 102239, ISSN 0141-9382, https://doi.org/10.1016/j.displa.2022.102239, (https://www.sciencedirect.com/science/article/pii/S014193822200066X)
26. Huff DT, Weisman AJ, Jeraj R. Interpretation and visualization techniques for deep learning models in medical imaging. Phys Med Biol. 2021 Feb 2;66(4):04TR01. https://doi.org/10.1088/1361-6560/abcd17. PMID: 33227719; PMCID: PMC8236074.
27. Carvalho LE, Sobieranski AC, von Wangenheim A. 3D Segmentation Algorithms for Computerized Tomographic Imaging: a Systematic Literature Review. J Digit Imaging. 2018 Dec;31(6):799–850. https://doi.org/10.1007/s10278-018-0101-z. PMID: 29915942; PMCID: PMC6261188.

28. Mubashar, M., Ali, H., Grönlund, C. et al. R2U++: a multiscale recurrent residual U-Net with dense skip connections for medical image segmentation. Neural Comput & Applic 34, 17723–17739 (2022). https://doi.org/10.1007/s00521-022-07419-7
29. Liu, H., Hu, D., Li, H., Oguz, I. (2023). Medical Image Segmentation Using Deep Learning. In: Colliot, O. (eds) Machine Learning for Brain Disorders. Neuromethods, vol 197. Humana, New York, NY. https://doi.org/10.1007/978-1-0716-3195-9_13
30. Sarmah, M., Neelima, A. & Singh, H.R. Survey of methods and principles in three-dimensional reconstruction from two-dimensional medical images. Vis. Comput. Ind. Biomed. Art 6, 15 (2023). https://doi.org/10.1186/s42492-023-00142-7
31. Sahiner B, Pezeshk A, Hadjiiski LM, Wang X, Drukker K, Cha KH, Summers RM, Giger ML. Deep learning in medical imaging and radiation therapy. Med Phys. 2019 Jan;46(1):e1–e36. https://doi.org/10.1002/mp.13264. Epub 2018 Nov 20. PMID: 30367497; PMCID: PMC9560030.
32. Olaf Ronneberger, Philipp Fischer, Thomas Brox, U-Net: Convolutional Networks for Biomedical Image Segmentation, arXiv:1505.04597, https://doi.org/10.48550/arXiv.1505.04597., May 2015.
33. Huang, S.-Y.; Hsu, W.-L.; Hsu, R.-J.; Liu, D.-W. Fully Convolutional Network for the Semantic Segmentation of Medical Images: A Survey. Diagnostics 2022, 12, 2765. https://doi.org/10.3390/diagnostics12112765
34. Devakishan Adla, G. Venkata Rami Reddy, Padmalaya Nayak, G. Karuna, A full-resolution convolutional network with a dynamic graph cut algorithm for skin cancer classification and detection, Healthcare Analytics, Volume 3, 2023, 100154, ISSN 2772-4425, https://doi.org/10.1016/j.health.2023.100154, (https://www.sciencedirect.com/science/article/pii/S2772442523000217)
35. Diwan, T., Anirudh, G. & Tembhurne, J.V. Object detection using YOLO: challenges, architectural successors, datasets and applications. Multimed Tools Appl 82, 9243–9275 (2023). https://doi.org/10.1007/s11042-022-13644-y
36. Shou Y, Meng T, Ai W, Xie C, Liu H, Wang Y. Object Detection in Medical Images Based on Hierarchical Transformer and Mask Mechanism. Comput Intell Neurosci. 2022 Aug 4;2022:5863782. https://doi.org/10.1155/2022/5863782. PMID: 35965770; PMCID: PMC9371842.
37. Malhotra P, Gupta S, Koundal D, Zaguia A, Enbeyle W. Deep Neural Networks for Medical Image Segmentation. J Healthc Eng. 2022 Mar 10;2022:9580991. https://doi.org/10.1155/2022/9580991. Retraction in: J Healthc Eng. 2023 Oct 11;2023:9781975. PMID: 35310182; PMCID: PMC8930223.
38. Sahin, M.E., Ulutas, H., Yuce, E. et al. Detection and classification of COVID-19 by using faster R-CNN and mask R-CNN on CT images. Neural Comput & Applic 35, 13597–13611 (2023). https://doi.org/10.1007/s00521-023-08450-y
39. Kaur, J., Singh, W. Tools, techniques, datasets and application areas for object detection in an image: a review. Multimed Tools Appl 81, 38297–38351 (2022). https://doi.org/10.1007/s11042-022-13153-y
40. Ribli, D., Horváth, A., Unger, Z. et al. Detecting and classifying lesions in mammograms with Deep Learning. Sci Rep 8, 4165 (2018). https://doi.org/10.1038/s41598-018-22437-z
41. How single-shot detector (SSD) works? URL: https://developers.arcgis.com/python/guide/how-ssd-works/
42. Zijian Zhou*, Jeremiah W. Sanders*, Jason M. Johnson, Maria K. Gule-Monroe, Melissa M. Chen, Tina M. Briere, Yan Wang, Jong Bum Son, Mark D. Pagel, Jing Li, Jingfei Ma, Computer-aided Detection of Brain Metastases in T1-weighted MRI for Stereotactic Radiosurgery Using Deep Learning Single-Shot Detectors, Radiology, Mar 17 2020, https://doi.org/10.1148/radiol.2020191479
43. Wenzel, M. (2023). Generative Adversarial Networks and Other Generative Models. In: Colliot, O. (eds) Machine Learning for Brain Disorders. Neuromethods, vol 197. Humana, New York, NY. https://doi.org/10.1007/978-1-0716-3195-9_5

44. Saeed, A.Q., Sheikh Abdullah, S.N.H., Che-Hamzah, J. et al. Synthesizing Retinal Images using End-To-End VAEs-GAN Pipeline-Based Sharpening and Varying Layer. Multimed Tools Appl 83, 1283–1307 (2024). https://doi.org/10.1007/s11042-023-17058-2
45. Manal Alamir, Manal Alghamdi, The Role of Generative Adversarial Network in Medical Image Analysis: An In-depth Survey, ACM Computing Surveys, Volume 55, Issue 5, Article No.: 96, pp 1–36, https://doi.org/10.1145/3527849, 03 December 2022.
46. Koshino K, Werner RA, Pomper MG, Bundschuh RA, Toriumi F, Higuchi T, Rowe SP. Narrative review of generative adversarial networks in medical and molecular imaging. Ann Transl Med. 2021 May;9(9):821. https://doi.org/10.21037/atm-20-6325. PMID: 34268434; PMCID: PMC8246192.
47. Jan Ehrhardt, Matthias Wilms, Chapter 8 – Autoencoders and variational autoencoders in medical image analysis, Editor(s): Ninon Burgos, David Svoboda, In The MICCAI Society book Series, Biomedical Image Synthesis and Simulation, Academic Press, 2022, Pages 129–162, ISBN 9780128243497, https://doi.org/10.1016/B978-0-12-824349-7.00015-3., (https://www.sciencedirect.com/science/article/pii/B9780128243497000153).

Chapter 5
Image Processing and Analysis

5.1 Introduction

Medical images have become widespread since the discovery of X-rays in 1895 [1]. In recent years, particularly following the introduction of computed tomography (CT) in 1972, there has been a significant surge in the utilization of medical imaging, leading to a substantial increase in the volume of medical image data [2].

It is estimated that the global healthcare sector contributed over 2.3 zettabytes out of the total 64.2 zettabytes generated worldwide in 2020 [3]. Approximately 90% of all healthcare data consists of imaging data. Notably, this medical image data is primarily generated in radiology departments through X-rays, CT scans, and MRIs [4].

With the expanding usage of medical imaging and the continuous improvement in the size and resolution of medical images, medical image processing has emerged as a crucial subfield within image processing. This field is gaining prominence and capturing the interest of the scientific community.

"Image processing and analysis" in medical imaging entail fundamental techniques aimed at enhancing, clarifying, and extracting meaningful information from images. One key principle is image enhancement, akin to adjusting the brightness and contrast in a photo, which helps improve the clarity of X-rays or MRIs, aiding in precise identification of ailments. Restoration techniques, comparable to removing scratches from old photographs, aim to eliminate distortions or noise in scans, ensuring accurate interpretations. Segmentation, such as using different colors to highlight distinct organs in a body scan, enables precise identification of regions of interest for detailed analysis.

Moreover, feature extraction and classification play pivotal roles in medical imaging. Similar to measuring the density of tissues or quantifying the size of a tumor in a CT scan, feature extraction identifies critical attributes for diagnosis and monitoring. Image classification, comparable to distinguishing between objects in a

E. Seeram, V. Kanade, *Artificial Intelligence in Medical Imaging Technology*,
https://doi.org/10.1007/978-3-031-64049-0_5

picture, involves categorizing different tissues or anomalies based on their unique characteristics in scans. Additionally, image fusion techniques amalgamate multiple modalities, such as MRI, CT, or PET scans. They create comprehensive and layered views resembling overlaying elements in a drawing, thereby providing a holistic perspective crucial for accurate medical assessments.

All the above processes collectively form the cornerstone of image processing and analysis, thereby enabling healthcare professionals to derive crucial insights from visual data for effective diagnosis and treatment planning.

A typical image processing and analysis procedure follows a sequence of steps to obtain the desired results, as illustrated in Fig. 5.1.

The following section delves deeper into each of the image processing steps mentioned in the above flowchart that significantly influences clinical decisions. It is vital to consider that the initial stage of image processing and analysis commences with the acquisition of raw image data using diverse devices such as CT scanners or MRI machines. This is followed by pre-processing, segmentation, feature extraction, classification, post-processing, and interpretation and visualization [5].

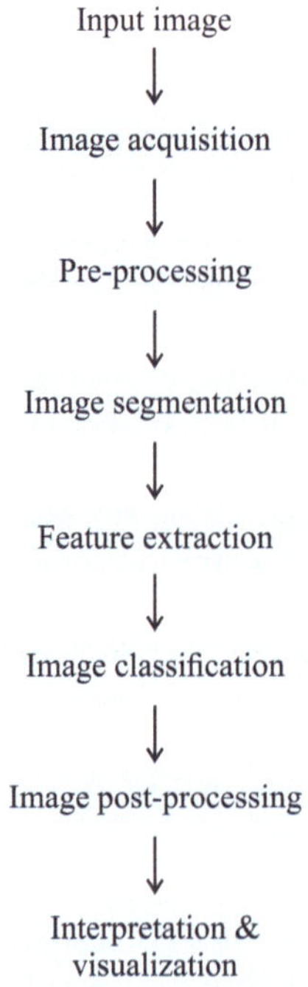

Fig. 5.1 Image processing and analysis flowchart. See text for further explanation

5.2 Pre-processing Techniques

After obtaining the raw medical image in step I, it goes through pre-processing.

Pre-processing techniques encompass various methods that prepare medical images for better interpretation and analysis. Here are some key techniques and their significance with examples:

1. **Noise Reduction**

Noise reduction in medical imaging holds immense significance due to its role in refining the quality and reliability of diagnostic information [6]. Medical images, whether from X-rays, MRIs, or ultrasounds, frequently encounter disruptive noise, whether from inherent equipment limitations or patient motion. This noise can obscure vital details, compromising the accuracy of diagnoses.

By employing noise reduction techniques similar to using filters to clarify a blurry photo, medical professionals can eliminate these disturbances, ensuring that images are clearer and more precise. This refinement is crucial as it enables radiologists and clinicians to discern subtle nuances within the images, facilitating more accurate diagnoses and effective treatment planning for patients.

For instance, in an ultrasound scan, speckle noise can distort images, making it harder to identify tissues accurately. Noise reduction techniques, similar to using noise reduction filters in photography apps, help clean up these images for clearer visualization of organs or abnormalities.

2. **Image Enhancement**

Image enhancement plays a critical role in refining the quality and interpretability of medical images, holding immense significance in diagnostic accuracy [6]. Often, medical images such as CT scans or ultrasounds may exhibit insufficient contrast or varying levels of brightness, hindering the visibility of critical details.

By employing image enhancement techniques similar to adjusting settings on a camera to bring out finer details, healthcare professionals can modify contrast, brightness, or sharpness to accentuate subtle elements within the images. This enhancement empowers radiologists and clinicians to discern minute variations in tissue density, pinpoint anomalies, or detect minute abnormalities that might otherwise go unnoticed. Ultimately, this refined quality in medical images aids in more precise diagnoses and facilitates effective and tailored treatment strategies for patients.

3. **Image Registration**

Image registration stands as a critical process in medical imaging, as it facilitates in comprehensive analysis and continual monitoring of a patient's condition. In medical practice, it's common to acquire images from various modalities or different time points to gather a comprehensive understanding of a patient's health status. Aligning these diverse images accurately—such as aligning MRI, CT, or PET scans—is essential for comparative analysis.

Moreover, image registration is used to correct for spatial misalignments, distortions, or differences in scale and orientation between images [7]. By registering images, practitioners can create a common reference frame, making it easier to extract meaningful information and enhance the accuracy of subsequent analyses.

For instance, in radiation therapy planning, combining CT and MRI scans requires precise alignment. Image registration techniques ensure accurate fusion of these images, assisting doctors in targeting tumors more effectively while sparing healthy tissues. This alignment, similar to aligning puzzle pieces to reveal a complete picture, ensures that imaging professionals can clearly compare and contrast images. It allows practitioners to make more informed decision, personalize treatment planning, and assist in ongoing monitoring of a patient's progress and response to interventions.

4. **Image Denoising and Deblurring**

Image denoising and deblurring techniques are vital in medical imaging as they contribute significantly to refining image clarity and overall quality. Medical images, whether from endoscopies, ultrasounds, or other modalities, often suffer from inherent blur or artifacts that can compromise diagnostic processes.

By employing denoising and deblurring methods, healthcare professionals can effectively mitigate distortions and artifacts in these images, enhancing their sharpness and clarity [8]. This refinement is crucial as it allows for clearer visualization of anatomical structures or pathological findings, enabling radiologists and clinicians to make more accurate assessments and diagnoses.

For instance, in endoscopy images, blurriness or artifacts might hinder the identification of lesions. Denoising and deblurring techniques can refine these images for better lesion identification, aiding in precise diagnosis.

These pre-processing techniques are essential as they ensure medical images are clearer, more accurate, and free from distortions or inconsistencies. Each preprocessing technique contributes significantly to medical imaging by refining raw images.

5.3 Segmentation Methods for Extracting Regions of Interest

In step III, the pre-processed image undergoes segmentation to identify and extract regions of interest for further processing.

Segmentation methods in medical imaging are like skilled painters meticulously isolating and highlighting specific elements in a complex canvas. They involve the precise extraction of areas or regions of interest from intricate medical images, providing critical information for diagnosis, treatment, and research [9].

For instance, in neuroimaging, segmenting brain images allows for the precise localization and measurement of tumors or lesions, aiding in surgical planning or monitoring disease progression. This segmentation process, similar to outlining

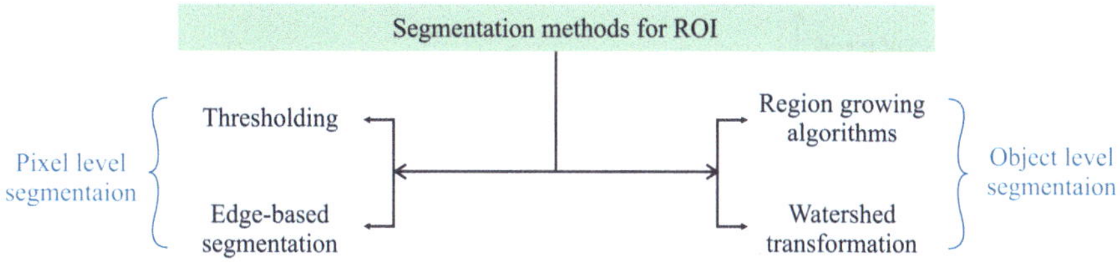

Fig. 5.2 Segmentation methods for extracting ROI. See text for further explanation

different sections on a map, provides professionals with a clearer understanding of the spatial distribution of anatomical features or pathological conditions.

The following section explores these methods with various examples, showcasing their importance in different imaging modalities (Fig. 5.2).

1. **Thresholding: Separating the Foreground from the Background**

Thresholding is like using a highlighter to mark important information. In a chest X-ray, it's crucial for separating bones from the surrounding soft tissues, just like drawing a clear line between sunlight and shadows. This technique works by distinguishing different levels of brightness in pixels, helping identify bones accurately and spotting any fractures or irregularities.

For instance, consider the concept of global thresholding, a technique assuming that an image possesses a bimodal histogram. This assumption suggests that by using a straightforward comparison operation between image values and a specific threshold value, marked as T, the extraction of an object from its background becomes feasible [10]. For instance, in an image represented by $f(x, y)$ with a histogram exhibiting two dominant modes, where the object and background pixels hold distinct gray levels, choosing a threshold value T effectively separates these modes (Fig. 5.3).

By applying this threshold value, denoted as T, the resulting thresholded image, $g(x, y)$, is created. This binary image encapsulates the essence of the object extraction process: pixels with an intensity value of 1 signify the presence of objects, while those assigned a value of 0 represent the background.

The thresholded image $g(x,y)$ is defined as

$$g(x,y) = \begin{cases} 1 \text{ if } (x,y) > T \\ 0 \text{ if } (x,y) \le T. \end{cases}$$

2. **Edge-Based Segmentation: Outlining Distinct Boundaries**

Edge-based segmentation is a method in image processing that can be likened to tracing the contours of a drawing with a sharp pencil. This technique is particularly applicable in medical imaging, such as in the context of a heart ultrasound. In the case of a heart ultrasound, edge-based segmentation involves pinpointing abrupt

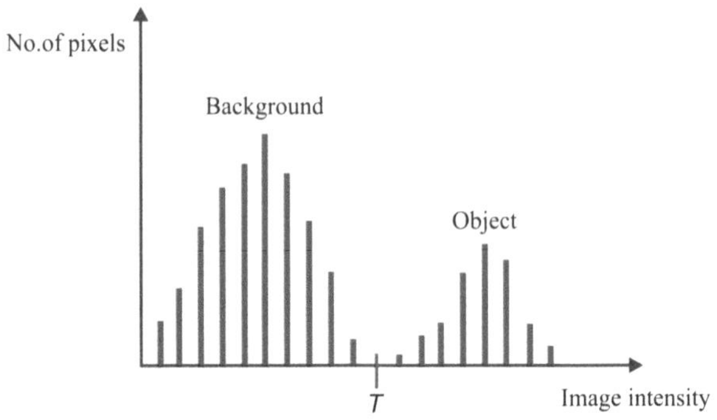

Fig. 5.3 Thresholding [10]. See text for further explanation

shifts in pixel brightness to create clear outlines around the heart's chambers [11]. It's like drawing distinct lines around specific shapes within the ultrasound image, allowing for a focused and precise identification of the anatomical structures within the heart. This approach aids in the early detection of any irregularities or anomalies in the heart, offering valuable insights for timely medical interventions.

3. **Region-Growing Algorithms: Grouping Similar Pixels**

Region-growing algorithms work like gathering items of the same color in a painting. In microscopic images, these algorithms detect and group pixels with similar characteristics, much like collecting objects of the same hue. They're vital for identifying and outlining individual cells, supporting tasks such as cell counting, studying cellular structures, and investigating microscopic diseases. These algorithms act as sorting tools, organizing pixels based on similarities and providing insights into cellular analysis and disease exploration at a minuscule level [12].

4. **Watershed Transformation: Separating Overlapping Structures**

The watershed transformation serves as a powerful analogy for its application in analyzing brain MRI images, functioning much like the process of untangling intertwined threads in a fabric. In the intricate landscape of the brain captured by MRI scans, the watershed transformation excels at discerning and outlining different brain regions, even when they overlap or intertwine [13, 14]. This is particularly crucial for the analysis of brain structure and the diagnosis of neurological conditions. By highlighting the distinct boundaries between different regions within the brain, the watershed transformation enables a more detailed examination, offering a clearer and more refined understanding of the complex neural architecture.

The series of images in Fig. 5.4, as presented by S. M. Kamrul Hasan and colleagues in the research paper titled "Watershed-Matching Algorithm: A New Pathway for Brain Tumor Segmentation," demonstrate the stepwise application of watershed segmentation on an initial MR T1 image [14]:

(a) **Input MR T1 image**: The original image used for analysis.

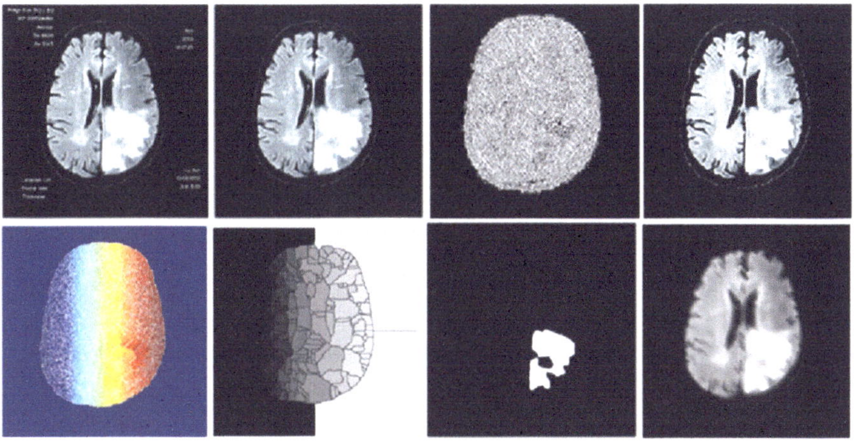

Fig. 5.4 Watershed segmentation [14]. See text for further explanation

(b) **Artifact removal, noise reduction, and edge smoothing**: Processed using trilateral filtering to enhance clarity by eliminating artifacts, reducing noise, and refining edges.
(c) **Denoising via bilateral filtering**: Further refinement achieved by employing a bilateral filter to enhance image quality and reduce noise.
(d) **Reduction of impulse noise using median filtering**: Employing median filtering to specifically target and minimize impulse noise.
(e) **Watershed transform in RGB color**: Application of the watershed transform in RGB color space to identify distinctive areas.
(f) **Gradient watershed transform**: Utilizing gradient-based watershed transformation for enhanced segmentation.
(g) **Tumor segmentation**: The segmented tumor area is highlighted in white within the final result.

This sequence illustrates the successive improvements and processing steps applied to the initial MR T1 image leading to the final segmentation of the tumor.

In essence, the watershed transformation plays a pivotal role in enhancing the interpretability of brain MRI images. Its ability to distinguish overlapping or intertwined brain regions ensures that the complexities of the neural anatomy are accurately represented and defined. This precision is instrumental in providing doctors with a comprehensive view, allowing them to diagnose neurological conditions with greater accuracy. It also aids in the development of targeted treatment plans tailored to the specific intricacies of the patient's brain structure.

5. **Deep Learning: Unraveling Complex Patterns**

Consider deep learning as a seasoned codebreaker solving complicated puzzles. In medical imaging, it employs neural networks to unravel and understand complex patterns for segmentation purposes. For instance, in CT scans, deep learning accurately identifies and delineates organs such as the liver, spleen, and kidneys [15]. Its precision in decoding these patterns is vital for treatment decisions and surgical procedures, providing detailed organ outlines that significantly enhance patient care.

The methods mentioned above can be categorized into two primary levels of segmentation: pixel-level (thresholding, edge-based segmentation) and object-level (region-growing algorithms, watershed transformation) [12, 14]. The following section offers insights into these broader level segmentation methods.

5.3.1 Pixel-Level and Object-Level Segmentation Approaches

Within the realm of medical image processing and analysis, the delineation of structures in an image is achieved through two distinctive strategies elaborated below:

Pixel-Level Segmentation
Pixel-level segmentation involves classifying each pixel in an image, treating segmentation as a pixel-wise classification problem. In medical imaging, this method is intensive but incredibly precise. For instance, in identifying tumors within an MRI scan, pixel-level segmentation precisely delineates each pixel constituting the tumor [16].

Techniques such as fully convolutional networks (FCNs) or U-Net are commonly used for pixel-level segmentation tasks in medical images. These algorithms learn to assign a label to each pixel, creating a detailed segmentation map that outline the tumor's boundaries or other areas of interest. This map helps doctors precisely locate and understand the extent of the disease (such as tumor), aiding in surgical planning or targeted therapies.

Object-Level Segmentation
Object-level segmentation focuses on identifying and delineating entire objects or structures within an image, rather than individual pixels [17]. Unlike pixel-level segmentation, this method segments whole organs or structures. For instance, in CT scans, object-level segmentation might aim to distinguish and delineate the liver, heart, or lungs as distinct objects within the image.

This segmentation provides a clear boundary outlining the object's (such as the liver) shape and position, assisting radiologists in quantitative analysis or pathology assessments. Algorithms such as mask R-CNN or region-based CNNs are utilized for object-level segmentation tasks in medical imaging.

Both pixel-level and object-level segmentation techniques play crucial roles in medical imaging, offering precise delineation of structures or abnormalities.

5.3.2 U-Net and Other Popular Architectures for Segmentation

As explained in Chap. 4, U-Net is a convolutional neural network specifically designed for medical image segmentation tasks. This architecture has a distinctive "U" shape, dividing into two main paths: one for understanding the bigger picture ("contracting path") and another for detailed identification ("expanding path").

The U-Net segmentation architecture is frequently utilized for various anatomical segmentation tasks derived from medical images. This includes but is not limited to applications such as segmenting organs like the liver, heart, and kidneys, as well as identifying structures such as tumors or abnormalities in medical imaging data [18].

Apart from U-Net, other popular segmentation architectures include the following:

1. **Fully convolutional network (FCN)**: Similar to U-Net but less interconnected. It focuses on recognizing individual pixels for segmentation without extensive connections (see Chap. 4). Its ability to handle input images of different sizes and produce pixel-wise predictions makes FCN well-suited for tasks that require detailed segmentation of structures such as organs, tissues, or abnormalities in images.
2. **V-Net**: V-Net is particularly effective in segmenting volumetric 3D medical images such as brain MRI or CT scans, accurately delineating structures in three-dimensional space [19].
3. **DeepLab**: DeepLab's atrous convolutions (a specific type of convolutional operation used in deep neural networks for image analysis tasks) are valuable in capturing multi-scale information, aiding in segmenting histopathology images where capturing fine details is crucial for diagnosis [20].
4. **Mask R-CNN**: Mask R-CNN is utilized for segmentation in medical images, accurately outlining and differentiating multiple instances of abnormalities or organs within an image [21].

Each architecture has its strengths, catering to different complexities and requirements in medical image segmentation. They offer diverse tools for precise and accurate segmentation crucial in medical diagnostics and treatment planning.

In summary, segmentation methods in medical imaging are the skilled tools that allow clinicians and researchers to dissect and analyze intricate details within images. Whether isolating bones from soft tissues in X-rays, delineating cardiac structures in ultrasounds, or precisely identifying organs in CT scans, these techniques are the artist's brushstrokes that unveil vital information critical for advancing medical diagnoses, treatment strategies, and scientific understanding.

5.4 Feature Extraction and Representation of Medical Images

Once the regions of interest are identified, step IV involves extracting typical features or patterns for analysis.

Feature extraction is similar to identifying distinct signatures or individual elements within images, much like spotting unique symbols in a coded message. Imagine an ultrasound image of the liver—feature extraction involves isolating and capturing specific patterns that represent healthy tissue versus abnormal growths or lesions [22]. Just as a cryptographer deciphers hidden symbols, feature extraction decodes these crucial indicators, aiding doctors in detecting liver abnormalities and tailoring treatment plans.

The following section explores this process further using various examples, showcasing its significance in medical imaging.

1. **Identifying Distinctive Patterns**

Consider examining a mammogram image of the breast. Feature extraction here involves pinpointing and emphasizing distinct tissue densities—similar to recognizing the differences between regular breast tissue and possible masses or calcifications. Just as an expert jeweler identifies specific facets in a diamond, feature extraction unveils these unique patterns within the image. This process assists radiologists in detecting abnormalities early and accurately characterizing irregularities within breast tissue.

2. **Texture Analysis for Anomaly Detection**

Imagine applying texture analysis to images of skin lesions. This method works like recognizing various textures in fabrics, but here, it identifies specific textures within skin lesions that indicate different conditions. For example, it can differentiate between normal and irregular textures in moles. This ability aids dermatologists in spotting potential melanomas early by analyzing these distinct patterns.

3. **Shape and Morphological Features**

Shapes and structures within medical images play a crucial role in understanding underlying conditions. Think about analyzing shapes in dental X-rays. Shape analysis, for instance, captures and represents the shapes and structures of teeth. Just as an architect recognizes unique building elements, these features help dentists spot abnormalities such as cavities or misalignments. This detailed information allows for precise dental treatments and thorough assessments.

4. **Intensity-Based Features for Diagnosis**

Intensities of pixels within images hold significant diagnostic information [23]. In X-rays, the variation in pixel intensities can indicate bone density or presence of fractures. Feature extraction here involves representing these intensities—highlighting differences between healthy and pathological tissues. This technique aids in identifying abnormalities such as fractures or osteoporosis, similar to recognizing variations in shades of color within a painting.

Pixel intensity analysis in retinal scans offers yet another illustration. This technique helps detect variations in pixel intensities related to conditions such as diabetic retinopathy. Thus, feature extraction in retinal scans aids ophthalmologists in early disease detection and management.

In essence, feature extraction in medical imaging decodes specific details within images, serving as unique indicators essential for accurate diagnoses and targeted medical treatments. Whether analyzing textures in skin lesions, capturing shapes in X-rays, or deciphering pixel intensities in retinal scans, these extracted features reveal vital details that guide medical fraternity toward informed decisions, ensuring better patient care.

5.5 Image Classification

In step V, the extracted features from the images are classified into various categories.

Medical image classification involves analyzing and categorizing various types of medical images, such as X-rays, MRI scans, CT scans, and ultrasound images, based on certain criteria [24]. The goal of this process is to distinguish among different anatomical structures, identify the presence or absence of specific diseases or abnormalities, and assist in the diagnostics by providing more precise and reliable information to medical professionals.

For instance, in the realm of oncology, medical image classification can streamline the identification of different stages of cancer by analyzing the shape, size, and texture of tumors in imaging scans, thus enhancing personalized patient care and treatment.

The following section delves further into the range of methods used to classify medical images.

5.5.1 Techniques for Categorizing Medical Images into Classes

Medical images can be categorized using one of the following methods, depending on the type of data associated with the medical image and the desired outcome:

1. **Binary classification**: For instance, the process of distinguishing normal from abnormal findings in a chest X-ray, such as identifying pneumonia, relies on binary classification. The CNN learns normal lung structures and contrasts them with pneumonia features for accurate classification [25] (Fig. 5.5).

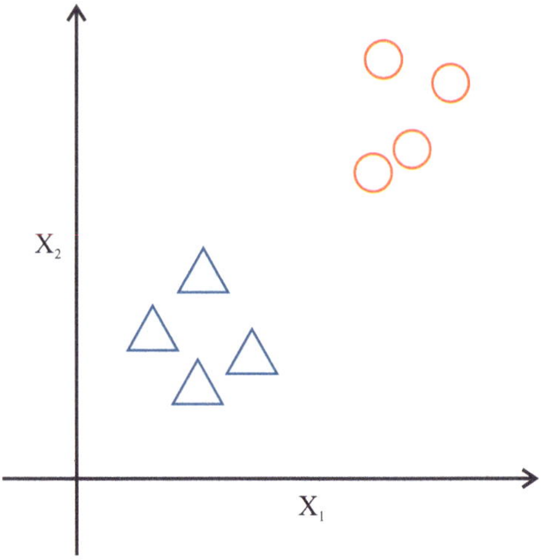

Fig. 5.5 Binary classification. See text for further explanation

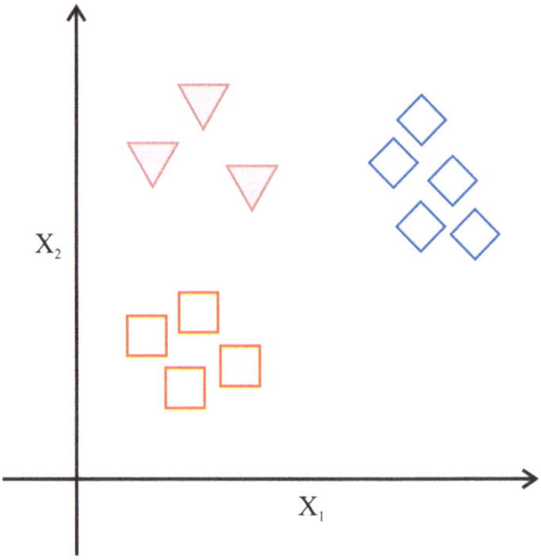

Fig. 5.6 Multi-class classification. See text for further explanation

The graph displayed depicts a binary classification scenario aimed at categorizing input X-rays into two distinct classes or categories. Each category is represented by a unique graphical structure, such as red squares and green circles, with the goal of sorting the X-rays into these defined groups.

2. **Multi-class classification**: In the analysis of brain MRI images, the classification of different types of brain tumors, such as gliomas and meningiomas, can be achieved through a multi-class classification method. This involves categorizing tumors based on their appearance and distinctive traits. The CNN learns patterns specific to each tumor type for accurate classification [26] (Fig. 5.6).

The graph depicted above showcases a scenario with multi-class scenario. The goal is to categorize the input data, which consists of MRI scans in this context, into multiple distinct classes. This involves distinguishing among vari-

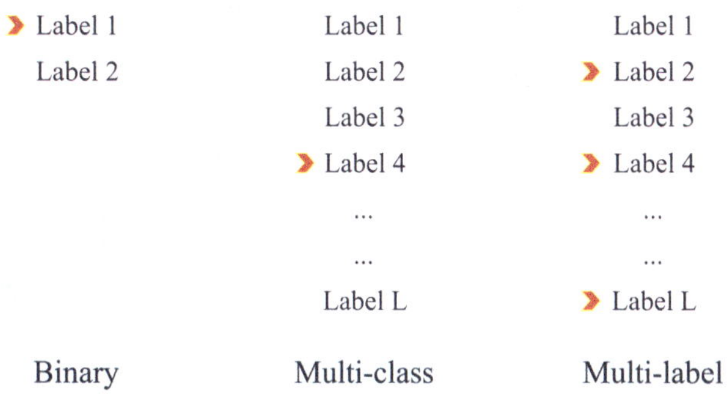

Fig. 5.7 Multi-label classification. See text for further explanation

ous types of brain tumors, such as medulloblastoma, oligodendroglioma, pituitary adenoma, ependymoma, or craniopharyngioma, as represented in the dataset.

3. **Multi-label Classification**: In a chest X-ray with multiple potential abnormalities, such as nodules, calcifications, and fractures, the CNN is trained to recognize and classify each condition independently [27].

   As illustrated in Fig. 5.7, in multi-label classification, unlike binary and multi-class classification, the deep learning model estimates the likelihood or probability of multiple classes simultaneously.

These techniques enable algorithms to identify critical patterns within medical images, facilitating accurate categorization into different diagnoses. Deep learning models adapt to diverse medical imaging scenarios, aiding precise diagnostics and targeted treatments based on image-based assessments.

5.5.2 *Handling Imbalanced Datasets and Data Augmentation*

In medical imaging, it's often common to have imbalanced datasets because some diseases or conditions are rarer than others. This can lead to a model being biased toward the majority class and performing poorly on the minority class [28].

To tackle this issue, there's a method called one-class classification that concentrates on building a model using samples from only one specific class. To help balance the data and improve learning, techniques such as data augmentation come into play. These methods involve making variations in the existing data, such as resizing, adding noise, flipping, tilting, and turning images, to create more training examples. Research has indicated that using these techniques tends to enhance how well deep learning models perform, especially when they are tested on new, unseen datasets.

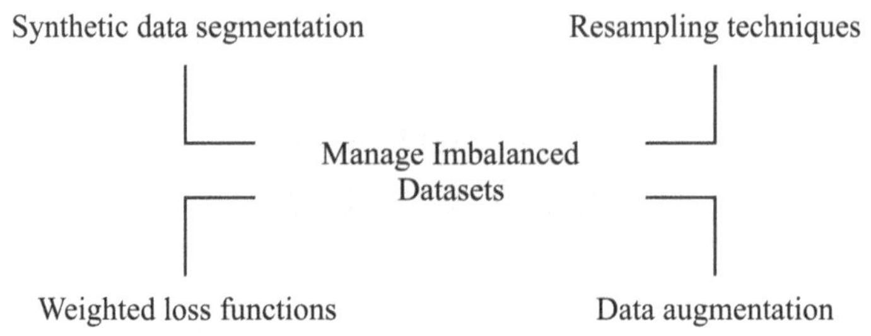

Fig. 5.8 Manage imbalanced datasets. See text for further explanation

The following section explores the approaches in-depth for managing imbalanced datasets in the field of medical imaging (Fig. 5.8):

Strategies for Handling Imbalanced Datasets
Some of the key strategies used to manage imbalanced datasets are as follows:

1. **Resampling Techniques**

Resampling techniques are strategies used to address the imbalance in class distribution within a dataset. These techniques aim to balance the representation of different classes for better model training [29].

- **Over-sampling the minority class**: This technique involves duplicating or creating more instances of the minority class, which is the less represented or rarer class in the dataset. By increasing the number of instances in the minority class, the dataset becomes more balanced, and the model gets exposed to a sufficient number of samples to learn from.
- **Under-sampling the majority class**: Under-sampling aims to reduce the number of instances in the majority class, which is the more frequent class in the dataset. By decreasing the number of instances in the majority class, the dataset becomes more balanced, allowing the model to give adequate attention to the minority class without being biased toward the majority class due to its larger representation.

2. **Synthetic Data Generation**

Synthetic data generation techniques, such as the synthetic minority over-sampling technique (SMOTE), focus on addressing imbalanced datasets by creating synthetic or artificial samples of the minority class:

- **Synthetic minority over-sampling technique (SMOTE)**: This method generates synthetic samples for the underrepresented or minority class by creating new instances that lie between existing minority class samples. It works by interpolating between neighboring instances in the feature space. For example, in a dataset with fewer instances of a rare disease, SMOTE generates new synthetic examples by creating combinations of nearby instances belonging to the same class [30].

3. **Weighted Loss Functions**

When dealing with imbalanced datasets, weighted loss functions are used during the training of machine learning models. The goal is to address the issue of underrepresented classes by assigning higher importance or penalties to errors made on the minority class samples [31].

Here's how it works:

(a) **Assigning higher weights**: By giving more weight to the errors made on the minority class (the rarer condition or less represented class), the model is encouraged to pay more attention to learning from these instances. This helps in ensuring that the model doesn't overlook or misclassify the less frequent class due to its scarcity in the dataset.
(b) **Penalizing minority class errors**: Errors made on the minority class samples are penalized more heavily compared to errors on the majority class. This encourages the model to focus on correctly identifying and learning the features specific to the underrepresented class.

4. **Data Augmentation**

Data augmentation is a technique used to expand the diversity and amount of data available for training by creating altered versions of existing images. The goal is to enhance the learning process of machine learning models, especially when working with limited datasets or imbalanced classes [32].

Here are some common techniques used in data augmentation:

- **Rotation**: Rotating images at different angles (like 90 or 180°) helps provide variations that the model can learn from. For instance, rotating chest X-ray images showing rare lung conditions can aid in exposing the model to diverse views of these conditions, improving its ability to recognize them from different perspectives.
- **Flipping**: Horizontal or vertical flipping of images creates mirrored versions, adding more diversity to the dataset. This is particularly useful in scenarios where the orientation of the subject in the image doesn't affect its interpretation.
- **Zooming**: Adjusting the zoom level in images helps the model understand the subject at different scales. This can be crucial in medical imaging, where details may vary based on the scale of observation.
- **Cropping**: Removing parts of an image or focusing on specific areas helps the model learn to identify relevant features. In medical imaging, cropping might be employed to emphasize certain structures or regions of interest.
- **Translation**: Shifting an image within its frame helps the model learn from variations in position or alignment. This technique can aid in simulating different viewpoints of an object or condition.

- **Adjusting brightness and contrast**: Altering the brightness or contrast of an image exposes the model to different lighting conditions, making it more robust to variations in image quality.

By applying these augmentation techniques, the dataset becomes more diverse, allowing the model to generalize better and improve its ability to recognize and understand various features in the images.

Example Use Case of Imbalanced Datasets
Imagine a dataset with 10,000 chest X-rays where only 500 indicate a rare lung condition. To address the imbalance, one could employ data augmentation techniques such as rotation, flipping, or altering brightness and contrast to create additional variations of the images representing the rare lung condition. Additionally, the use of weighted loss functions could prioritize the correct classification of these rare cases during the model's training.

These strategies ensure that the machine learning models trained on imbalanced medical image datasets are more capable of recognizing and accurately predicting rare conditions. It enhances the overall performance and reliability of the classification process.

5.6 Image Post-processing

Finally, in step VI, the categorized medical images undergo post-processing to enhance the overall quality of the processed images [33].

In the context of medical images, post-processing refers to the application of additional image manipulation and refinement techniques after the initial segmentation, feature extraction, and image classification. This step aims to enhance the visual quality, clarify anatomical details, and improve diagnostic information.

Common post-processing operations include noise reduction, fine-tuning measurements, selective filtering, contrast adjustment, image filtering, and three-dimensional reconstruction. These techniques help optimize medical images for better interpretation by professionals, ensuring accurate diagnosis and effective communication of critical information for patient care.

However, apart from these basic post-processing techniques, some advanced operations include the following:

In the context of medical image analysis, beyond the fundamental post-processing techniques, there are advanced operations designed to enhance the quality and utility of medical images (Fig. 5.9):

1. **Super-resolution imaging**: Super-resolution techniques go beyond the original resolution limits, improving the clarity and detail of medical images. By reconstructing high-resolution images from lower-resolution ones, these techniques are valuable for identifying finer structures or abnormalities in medical imaging.

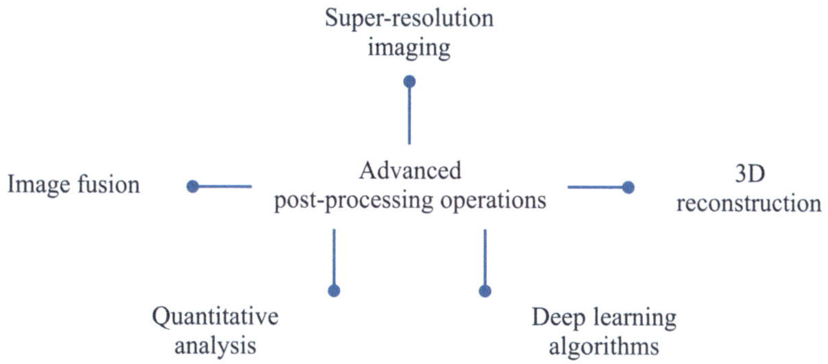

Fig. 5.9 Advanced post-processing operations. See text for further explanation

This enhancement can be particularly critical for precise diagnosis and treatment planning [34].

2. **Image fusion**: Image fusion involves the amalgamation of multiple images from different modalities or imaging sequences (e.g., combining MRI and CT scans). This process creates a comprehensive image, providing a more holistic view that can significantly aid in the diagnosis and planning of medical treatments by offering a complete understanding of the patient's condition [35].
3. **3D reconstruction**: Transforming 2D medical images into 3D models or volumes allows for improved visualization and understanding of complex anatomical structures. This is especially beneficial for surgeons in preoperative planning and educational contexts, providing a more immersive and detailed perspective [36].
4. **Quantitative analysis**: Leveraging algorithms for measuring aspects such as volume, density, or other quantitative parameters within medical images enables precise and objective analysis. This quantitative data plays a crucial role in on-point diagnosis and ongoing monitoring of treatment effectiveness [37].
5. **Deep learning algorithms**: Advanced AI algorithms, such as those used for image segmentation, object detection, and disease classification, automate and enhance the accuracy of medical image analysis. These methods, discussed extensively in various sections, contribute to the efficiency and reliability of tasks crucial in medical diagnostics and research [24].

5.6.1 Image Interpretation

Following the post-processing of images, the focus shifts to image interpretation, visualization, and presentation. This is crucial for making well-informed clinical decisions by analyzing the processed images [38].

Image interpretation refers to the careful analysis and understanding of the processed images. The goal is to extract meaningful information from these images to help healthcare professionals make informed decisions about a patient's health.

The process begins with visual analysis where the healthcare professionals, such as radiologists or doctors, visually examine the processed medical images to identify structures, organs, and any abnormalities. They use their knowledge of anatomy and pathology to recognize patterns and detect variations in the images.

Next, the doctors tend to build a mental model of the patient's internal anatomy based on the observed features in the image. This mental model serves as a reference for understanding what is normal and what might indicate a medical condition. Then, the current images are compared with any previous images of the same patient (if available) or with standard reference images. Here, differences or changes over time can be crucial indicators of disease progression or response to treatment.

In addition to such visual inspection, quantitative analysis may be performed using computer-aided tools. These tools can measure sizes, densities, and other characteristics, providing additional data for a more comprehensive evaluation.

Common analytical methods applied in image analysis include filtered back projection (FBP) and the Fourier transform (FT). FBP is widely used in tomography, while FT is particularly applied in MRI. These methods facilitate the transformation of image data into a visual format that can be easily interpreted by medical professionals [39, 40].

5.6.2 *Visualization and Presentation*

Subsequently, the results derived from image interpretation are translated into a visually comprehensible format through the process of visualization and presentation. This step involves the use of specialized tools and techniques to effectively communicate the findings to various stakeholders, including fellow healthcare professionals, patients, and decision-makers.

Visualization tools may include three-dimensional reconstructions, color-coded maps, or annotated overlays that highlight specific regions of interest [41]. Furthermore, methods such as multi-planar reformation (MPR) and maximum intensity projection (MIP) are utilized to visualize medical images across various planes, revealing areas or regions of particular importance [42, 43].

Such visual representation plays a pivotal role in the collaborative decision-making process, enabling informed discussions and, ultimately, enhancing the quality of patient care.

References

1. Babic RR, Stankovic Babic G, Babic SR, Babic NR. 120 YEARS SINCE THE DISCOVERY OF X-RAYS. Med Pregl. 2016 Sep;69(9–10):323–330. https://doi.org/10.2298/mpns1610323b.
2. Hermena S, Young M. CT-scan Image Production Procedures. [Updated 2023 Aug 8]. In: StatPearls [Internet]. Treasure Island (FL): StatPearls Publishing; 2024 Jan-. Available from: https://www.ncbi.nlm.nih.gov/books/NBK574548/
3. Total amount of global healthcare data generated in 2013 and a projection for 2020*(in exabytes), Statista, URL: https://www.statista.com/statistics/1037970/global-healthcare-data-volume/
4. S. K. Zhou et al., "A Review of Deep Learning in Medical Imaging: Imaging Traits, Technology Trends, Case Studies With Progress Highlights, and Future Promises," in Proceedings of the IEEE, vol. 109, no. 5, pp. 820–838, May 2021, https://doi.org/10.1109/JPROC.2021.3054390.
5. Deserno, T.M. (2010). Fundamentals of Biomedical Image Processing. In: Deserno, T. (eds) Biomedical Image Processing. Biological and Medical Physics, Biomedical Engineering. Springer, Berlin, Heidelberg. https://doi.org/10.1007/978-3-642-15816-2_1
6. Chen, Z., Pawar, K., Ekanayake, M. et al. Deep Learning for Image Enhancement and Correction in Magnetic Resonance Imaging—State-of-the-Art and Challenges. J Digit Imaging 36, 204–230 (2023). https://doi.org/10.1007/s10278-022-00721-9
7. Chen, M., Tustison, N.J., Jena, R., Gee, J.C. (2023). Image Registration: Fundamentals and Recent Advances Based on Deep Learning. In: Colliot, O. (eds) Machine Learning for Brain Disorders. Neuromethods, vol 197. Humana, New York, NY. https://doi.org/10.1007/978-1-0716-3195-9_14
8. Cammarasana, S., Nicolardi, P. & Patanè, G. Real-time denoising of ultrasound images based on deep learning. Med Biol Eng Comput 60, 2229–2244 (2022). https://doi.org/10.1007/s11517-022-02573-5
9. Huang, C., Liu, Q. & Yu, S. Regions of interest extraction from color image based on visual saliency. J Supercomput 58, 20–33 (2011). https://doi.org/10.1007/s11227-010-0532-x
10. Jadwiga Rogowska, in Handbook of Medical Image Processing and Analysis (Second Edition), 2009, Overview and Fundamentals of Medical Image Segmentation, https://www.sciencedirect.com/topics/engineering/global-thresholding
11. Cao, X., Ding, W., Hu, S., Su, L. (2013). Image Segmentation Based on Edge Growth. In: Lu, W., Cai, G., Liu, W., Xing, W. (eds) Proceedings of the 2012 International Conference on Information Technology and Software Engineering. Lecture Notes in Electrical Engineering, vol 212. Springer, Berlin, Heidelberg. https://doi.org/10.1007/978-3-642-34531-9_57
12. Shih, F.Y. (2009). Image Segmentation. In: LIU, L., ÖZSU, M.T. (eds) Encyclopedia of Database Systems. Springer, Boston, MA. https://doi.org/10.1007/978-0-387-39940-9_1011
13. Puranik, M.M., Krishnan, S. (2011). Segmentation of Image Using Watershed and Fast Level Set Methods. In: Das, V.V., Thomas, G., Lumban Gaol, F. (eds) Information Technology and Mobile Communication. AIM 2011. Communications in Computer and Information Science, vol 147. Springer, Berlin, Heidelberg. https://doi.org/10.1007/978-3-642-20573-6_40
14. Hasan, S.M.K., Sarkar, Y., Ahmad, M. (2018). Watershed-Matching Algorithm: A New Pathway for Brain Tumor Segmentation. In: Kalam, A., Das, S., Sharma, K. (eds) Advances in Electronics, Communication and Computing. Lecture Notes in Electrical Engineering, vol 443. Springer, Singapore. https://doi.org/10.1007/978-981-10-4765-7_5
15. Schoppe O, Pan C, Coronel J, Mai H, Rong Z, Todorov MI, Müskes A, Navarro F, Li H, Ertürk A, Menze BH. Deep learning-enabled multi-organ segmentation in whole-body mouse scans. Nat Commun. 2020 Nov 6;11(1):5626. https://doi.org/10.1038/s41467-020-19449-7. PMID: 33159057; PMCID: PMC7648799.
16. Lee, H., Troschel, F.M., Tajmir, S. et al. Pixel-Level Deep Segmentation: Artificial Intelligence Quantifies Muscle on Computed Tomography for Body Morphometric Analysis. J Digit Imaging 30, 487–498 (2017). https://doi.org/10.1007/s10278-017-9988-z
17. Sharma, R., Saqib, M., Lin, C.T. et al. A Survey on Object Instance Segmentation. SN COMPUT. SCI. 3, 499 (2022). https://doi.org/10.1007/s42979-022-01407-3

18. Reza Azad, et al., Medical Image Segmentation Review: The success of U-Net, arXiv:2211.14830
19. Hesamian, M.H., Jia, W., He, X. et al. Deep Learning Techniques for Medical Image Segmentation: Achievements and Challenges. J Digit Imaging 32, 582–596 (2019). https://doi.org/10.1007/s10278-019-00227-x
20. Liang-Chieh Chen, et al., DeepLab: Semantic Image Segmentation with Deep Convolutional Nets, Atrous Convolution, and Fully Connected CRFs, May 2017, URL: https://arxiv.org/pdf/1606.00915.pdf
21. Iqbal, S., A. N. Qureshi, Li, J. et al. On the Analyses of Medical Images Using Traditional Machine Learning Techniques and Convolutional Neural Networks. Arch Computat Methods Eng 30, 3173–3233 (2023). https://doi.org/10.1007/s11831-023-09899-9
22. Puttagunta, M., Ravi, S. Medical image analysis based on deep learning approach. Multimed Tools Appl 80, 24365–24398 (2021). https://doi.org/10.1007/s11042-021-10707-4
23. Archana, R., Jeevaraj, P.S.E. Deep learning models for digital image processing: a review. Artif Intell Rev 57, 11 (2024). https://doi.org/10.1007/s10462-023-10631-z
24. Li M, Jiang Y, Zhang Y, Zhu H. Medical image analysis using deep learning algorithms. Front Public Health. 2023 Nov 7;11:1273253. https://doi.org/10.3389/fpubh.2023.1273253. PMID: 38026291; PMCID: PMC10662291.
25. Goyal, S., Singh, R. Detection and classification of lung diseases for pneumonia and Covid-19 using machine and deep learning techniques. J Ambient Intell Human Comput 14, 3239–3259 (2023). https://doi.org/10.1007/s12652-021-03464-7
26. Irmak, E. Multi-Classification of Brain Tumor MRI Images Using Deep Convolutional Neural Network with Fully Optimized Framework. Iran J Sci Technol Trans Electr Eng 45, 1015–1036 (2021). https://doi.org/10.1007/s40998-021-00426-9
27. Kufel J, Bielówka M, Rojek M, Mitręga A, Lewandowski P, Cebula M, Krawczyk D, Bielówka M, Kondoł D, Bargieł-Łączek K, Paszkiewicz I, Czogalik Ł, Kaczyńska D, Wocław A, Gruszczyńska K, Nawrat Z. Multi-Label Classification of Chest X-ray Abnormalities Using Transfer Learning Techniques. J Pers Med. 2023 Sep 22;13(10):1426. https://doi.org/10.3390/jpm13101426. PMID: 37888037; PMCID: PMC10607847.
28. Gao L, Zhang L, Liu C, Wu S. Handling imbalanced medical image data: A deep-learning-based one-class classification approach. Artif Intell Med. 2020 Aug;108:101935. https://doi.org/10.1016/j.artmed.2020.101935. Epub 2020 Aug 7. PMID: 32972664; PMCID: PMC7519174.
29. Varotto G, Susi G, Tassi L, Gozzo F, Franceschetti S and Panzica F (2021) Comparison of Resampling Techniques for Imbalanced Datasets in Machine Learning: Application to Epileptogenic Zone Localization From Interictal Intracranial EEG Recordings in Patients With Focal Epilepsy. Front. Neuroinform. 15:715421. https://doi.org/10.3389/fninf.2021.715421
30. Sowjanya, A.M., Mrudula, O. Effective treatment of imbalanced datasets in health care using modified SMOTE coupled with stacked deep learning algorithms. Appl Nanosci 13, 1829–1840 (2023). https://doi.org/10.1007/s13204-021-02063-4
31. Iqbal S, Qureshi AN, Li J, Choudhry IA, Mahmood T. Dynamic learning for imbalanced data in learning chest X-ray and CT images. Heliyon. 2023 Jun 1;9(6):e16807. https://doi.org/10.1016/j.heliyon.2023.e16807. PMID: 37313141; PMCID: PMC10258426.
32. Goceri, E. Medical image data augmentation: techniques, comparisons and interpretations. Artif Intell Rev 56, 12561–12605 (2023). https://doi.org/10.1007/s10462-023-10453-z
33. Euclid Seeram, Image Postprocessing in Digital Radiology—A Primer for Technologists, Journal of Medical Imaging and Radiation Sciences, VOLUME 39, ISSUE 1, P23–41, MARCH 2008, https://doi.org/10.1016/j.jmir.2008.01.004
34. Umirzakova, S.; Mardieva, S.; Muksimova, S.; Ahmad, S.; Whangbo, T. Enhancing the Super-Resolution of Medical Images: Introducing the Deep Residual Feature Distillation Channel Attention Network for Optimized Performance and Efficiency. Bioengineering 2023, 10, 1332. https://doi.org/10.3390/bioengineering10111332

35. A.P. James, B. V. Dasarathy, Medical Image Fusion: A survey of the state of the art, Information Fusion, 2014.
36. Jordan Fletcher, Danilo Miskovic, Computer-Generated modelling in surgery, Seminars in Colon and Rectal Surgery, 2024, 101003, ISSN 1043-1489, https://doi.org/10.1016/j.scrs.2024.101003, (https://www.sciencedirect.com/science/article/pii/S1043148924000022)
37. Rea, G.; Sverzellati, N.; Bocchino, M.; Lieto, R.; Milanese, G.; D'Alto, M.; Bocchini, G.; Maniscalco, M.; Valente, T.; Sica, G. Beyond Visual Interpretation: Quantitative Analysis and Artificial Intelligence in Interstitial Lung Disease Diagnosis "Expanding Horizons in Radiology". Diagnostics 2023, 13, 2333. https://doi.org/10.3390/diagnostics13142333
38. Huff DT, Weisman AJ, Jeraj R. Interpretation and visualization techniques for deep learning models in medical imaging. Phys Med Biol. 2021 Feb 2;66(4):04TR01. https://doi.org/10.1088/1361-6560/abcd17. PMID: 33227719; PMCID: PMC8236074.
39. Ali Raza Barket, Weidong Hu, Bing Wang, Waseem Shahzad, and Jabir Shabbir Malik, "Selection criteria of image reconstruction algorithms for terahertz short-range imaging applications," Opt. Express 30, 23398–23416 (2022)
40. J. Fessler, Chapter 3, Analytical Tomographic Image Reconstruction Methods, October 25, 2021, URL: http://web.eecs.umich.edu/~fessler/book/c-tomo.pdf
41. Kather, J., Weidner, A., Attenberger, U. et al. Color-coded visualization of magnetic resonance imaging multiparametric maps. Sci Rep 7, 41107 (2017). https://doi.org/10.1038/srep41107
42. Isogai S, Takehara Y, Isoda H, Kaneko M. [Maximum intensity projection (MIP) and multiplanar reformation (MPR) for post-processing cholangiopancreatographic data set—clinical application and pitfalls]. Nihon Rinsho. 1998 Nov;56(11):2760–7. Japanese.
43. A. Blum, R. Gillet, A. Rauch, A. Urbaneja, H. Biouichi, G. Dodin, E. Germain, C. Lombard, P. Jaquet, M. Louis, L. Simon, P. Gondim Teixeira, 3D reconstructions, 4D imaging and postprocessing with CT in musculoskeletal disorders: Past, present and future, Diagnostic and Interventional Imaging, Volume 101, Issue 11, 2020, Pages 693–705, ISSN 2211-5684, https://doi.org/10.1016/j.diii.2020.09.008, (https://www.sciencedirect.com/science/article/pii/S2211568420302254)

Chapter 6
Artificial Intelligence Applications in Medical Imaging

6.1 Introduction

Medical imaging has a timeline (Fig. 6.1) that includes several notable discoveries that earned the pioneers the Nobel Prize for their significant contributions to the imaging sciences, from the discovery of X-rays by WC Roentgen in 1895, who was awarded the first Nobel Prize in Physics in 1901, to the development of computed tomography (CT) by Sir Godfrey Hounsfield and Allan Cormack who shared the Nobel Prize in Medicine in 1979. Other imaging modalities include nuclear medicine; diagnostic medical ultrasound; single-photon emission CT (SPECT/CT) and positron emission CT (PET/CT), which are both hybrid imaging modalities; and magnetic resonance imaging (MRI) which earned the pioneers Paul Lauterbur and Sir Peter Mansfield the Nobel Prize in Medicine in 2003. As illustrated in Fig. 6.1, in 1918, the Food and Drug Administration (FDA) in the United States began the approval of artificial intelligence (AI) algorithms for use in medical imaging [1].

The literature offers several papers on the applications of artificial intelligence (AI) in medical imaging [1–7]. These applications were listed in Chap. 1 and range from radiology workflow, computer-aided detection, and disease detection and classification to radiomics, imaging biobanks, dose optimization, and structured reporting. Furthermore, Kim et al. [5] identified digital image processing operations as yet another application of AI in medical imaging.

The purpose of this chapter is to describe these applications to an extent that meets the needs of imaging technologists/radiographers and other related professionals such as nurses, biomedical engineering technologists, and other related healthcare professionals interested in how AI is used in medical imaging.

E. Seeram, V. Kanade, *Artificial Intelligence in Medical Imaging Technology*,
https://doi.org/10.1007/978-3-031-64049-0_6

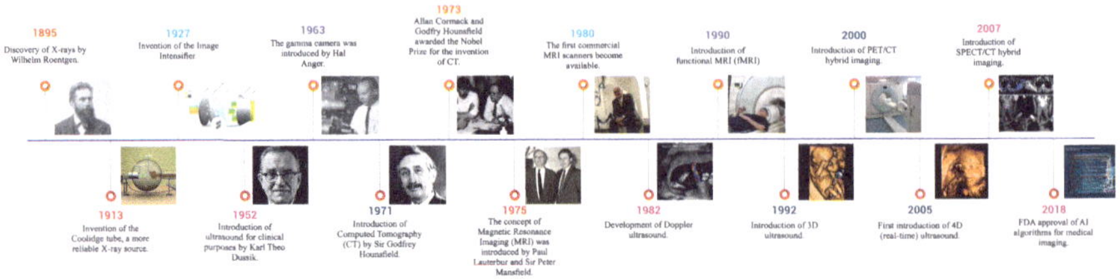

Fig. 6.1 A historical timeline of the key discoveries in medical imaging. (From Najjar, R. Redefining Radiology: A Review of Artificial Intelligence Integration in Medical Imaging. Diagnostics 2023, 13, 2760. https://doi.org/10.3390/diagnostics13172760. Reprinted under Creative Commons Attribution (CC BY) license (https://creativecommons.org/licenses/by/4))

6.2 Radiology Workflow

The term *workflow* literally means "the sequence of steps involved in moving from the beginning to the end of a working process" [8]. A Google search on January 20, 2024, using the term "radiology workflow," revealed 9,520,000 results. This information demonstrates the importance with which this topic has gained widespread attention in medical imaging. The fundamental steps in a radiology workflow paradigm include scheduling and preparation, image acquisition, image and data interpretation, outcomes and follow-up care, shared decision-making, pathway selection and treatment, reporting, and communication of the results [8].

The application of AI to radiology workflow has been discussed in the literature [1, 9–11]. For example, a streamlined workflow diagram illustrating the role of artificial intelligence in radiological practice [1] is illustrated in Fig. 6.2. The diagram identifies ten areas where AI can be applied to optimize workflow operations. These operations include scheduling, scanning, acquisition, interpretation, reporting, communication, recommendation, treatment response, adverse events, and follow-up and monitoring. For example, in scanning and interpretation (Fig. 6.2), AI ensures the right imaging procedure is selected, reducing radiation exposure by picking the optimal scan settings, and radiologists receive help from AI in interpreting images and spotting urgent cases, respectively. Furthermore, Pierre et al. [10] and Ranschaert et al. [11] showed that AI algorithms can provide improvements in workflow operations.

6.2.1 *ChatGPT and Radiology Workflow*

ChatGPT is an artificial intelligence (AI) chatbot that "uses natural language processing to create humanlike conversational dialogue [12, 13]. The language model can respond to questions and compose various written content, including articles, social media posts, essays, code, and emails" [14]. ChatGPT was created by OpenAI, an AI research and development company, whose mission is to "ensure that artificial general intelligence benefits all of humanity" [15].

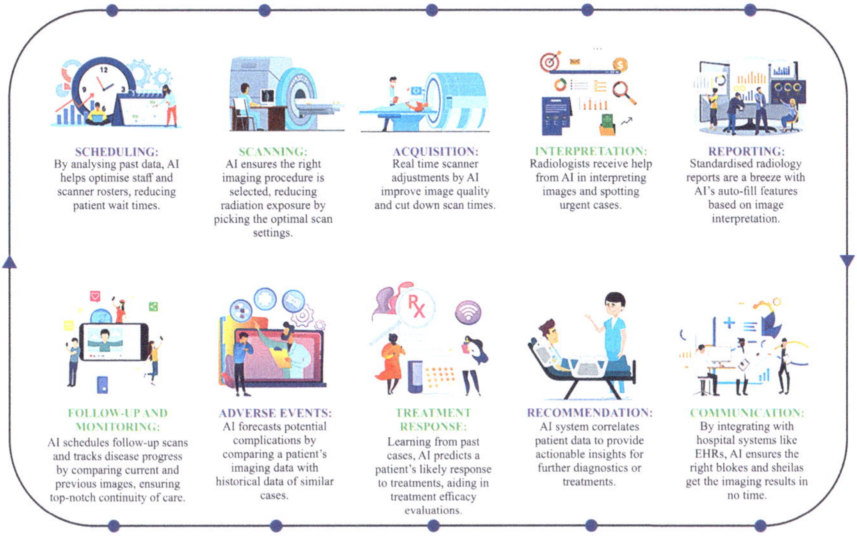

Fig. 6.2 A streamlined workflow diagram illustrating the role of artificial intelligence in radiological practice. (From Najjar, R. Redefining Radiology: A Review of Artificial Intelligence Integration in Medical Imaging. Diagnostics 2023, 13, 2760. https://doi.org/10.3390/diagnostics13172760. Reprinted under Creative Commons Attribution (CC BY) license (https://creativecommons.org/licenses/by/4))

Masse et al. [9] described how ChatGPT (based on deep learning algorithms) can be integrated into radiology workflow operations "to improve efficiency, accuracy, and patient care. ChatGPT can streamline various radiology workflow steps, including patient registration, scheduling, patient check-in, image acquisition, interpretation, and reporting." They report that "ChatGPT's AI technology boosts radiologists' diagnostic accuracy and efficiency." ChatGPT is now being applied in other areas of radiology, such as in oncology, and has been discussed in a paper entitled "Educating the next generation of radiologists: a comparative report of ChatGPT and e-learning resources."

The authors [9] also point out that the use of ChatGPT in this regard is not without shortcomings, noting the bias and ethical concerns. Furthermore, they also stress that as technology advances, "ChatGPT is likely to become an increasingly important tool in the field of radiology and in healthcare more broadly" [9].

6.3 Computer-Aided Detection/Diagnosis

The use of computers to assess mammographic images dates back to the late 1960s [16–18] and was initially intended to act as a second interpreter [19] (the radiologist being the first interpreter). Applications in this respect fall into two categories, namely, *computer-aided detection (CADe)* and *computer-aided diagnosis (CADx)* [19]. While the former is intended to reduce the possibility of missing lesions by identifying abnormal regions of the image; the latter is intended to provide assistance to the human observer in evaluating various pathologies, based on the principles of classification [19]. CADx was introduced in the 1990s as the next logical step in the development of these systems, with the goal of automating the classification of abnormalities [20].

With the advent of AI and its applications in medical imaging [21], CADx has moved from the lesion classification task for breast cancer, for example, into another dimension using deep learning technology (Chap. 4). CADe and CADx will be described further in Chap. 8. An extension of CAD is *radiomics* that "associates the quantitative characteristics (features) of images with patient data and clinical outcomes, not only allowing the diagnosis to be made but also providing information regarding the prognosis and treatment response" [22]. The essentials of radiomics will be described in Sect. 6.4.

6.4 Radiomics

The literature offers several definitions and reviews of radiomics that characterize the nature of this growing field of study in medicine [1, 2, 23, 24]. Radiomics was first defined in 2012 [25, 26] and later in 2015 [27]. These definitions reflect that *radiomics* is "the extraction of features from diagnostic images, the final product of which is a quantitative feature/parameter, measurable and mineable from images" [2]. Radiomics involves a number of processing operations such as pre-processing, segmentation, feature extraction, and feature analysis. These topics were described in detail in Chap. 5.

In a recent article, Haneberg et al. [7] reviewed the *traditional radiomic workflow*, as shown in Fig. 6.3, that illustrates four steps, namely, input image, lesion identification (detection, localization, and segmentation), feature extraction and prediction, and finally an outcome. The radiomic features extracted are subsequently "used as inputs for a statistical model or machine learning classifier to predict an outcome of interest" [7]. Furthermore, deep learning (subset of machine learning, as described in Chap. 1) algorithms can be used to perform not only image analysis but also classification tasks [7], as illustrated in Fig. 6.4, thus improving processing efficiency.

Radiomic analysis process involves a series of successive steps that elaborates its workflow, as illustrated in Fig. 6.5. These steps include image acquisition, image pre-processing, segmentation of the region of interest, feature extraction, feature selection, and model construction and evaluation [23, 24, 28].

It is not within the scope of this chapter to outline the full details of the radiomic pipeline; however, a brief summary of the steps as described by Scapicchio et al. [24] and McCague et al. [28] is as follows:

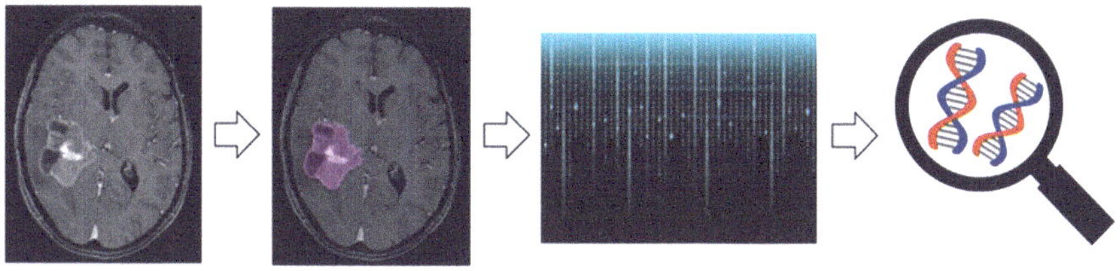

Fig. 6.3 Overview of the traditional radiomic workflow. See text for further explanation (From Haneberg AG, Pierre K, Winter-Reinhold E, Hochhegger B, Peters KR, Grajo J, Arreola M, Asadizanjani N, Bian J, Mancuso A, Forghani R (2023). Introduction to Radiomics and Artificial Intelligence: A Primer for Radiologists, Seminars in Roentgenology; 58: 2, 152–157. Reproduced by permission)

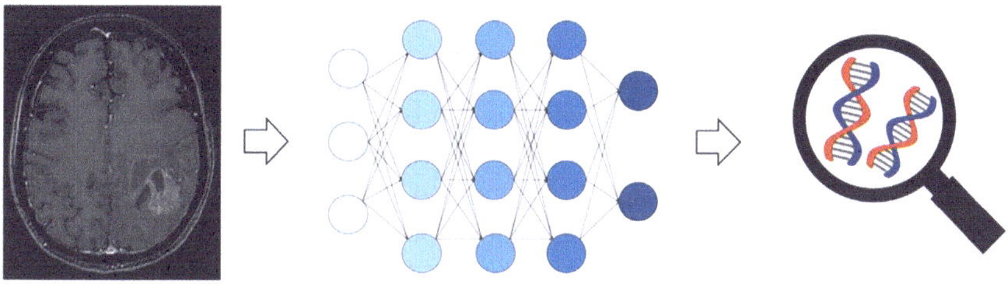

Fig. 6.4 The use of deep learning in the traditional radiomic workflow to perform image analysis and classification tasks, thus improving processing efficiency. (From Haneberg AG, Pierre K, Winter-Reinhold E, Hochhegger B, Peters KR, Grajo J, Arreola M, Asadizanjani N, Bian J, Mancuso A, Forghani R (2023). Introduction to Radiomics and Artificial Intelligence: A Primer for Radiologists, Seminars in Roentgenology; 58: 2, 152–157. Reproduced by permission)

- *Image acquisition*: This first step involves setting up a number of parameters depending on the modality (imaging or therapy) as well as tissue identification. Most studies have used CT and MRI images, such as hardware, acquisition factors for CT, and magnetic field strength and pulse sequences in MRI.
- *Image pre-processing*. This is the second step in which images are prepared for subsequent steps.

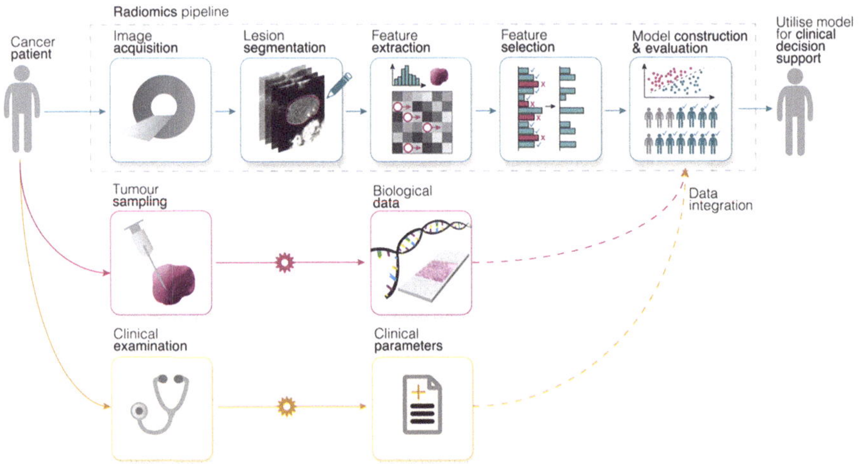

Fig. 6.5 Schematic illustration of a typical radiomic pipeline. See text for further explanation. (From McCague C, Ramlee S, Reinius M, Selby I, Hulse D, Piyatissa P, Bura V, Crispin-Ortuzar M, Sala E, R. Woitek R (2023). Introduction to radiomics for a clinical audience. Clinical Radiology; 78: (2). 83–98. Reprinted under Creative Commons Attribution license (https://creativecommons.org/licenses/by/4))

- *Image segmentation* of a region of interest (ROI) containing either normal tissue or a lesion represents the third step. This can be done automatically using segmentation software or manually by expert radiologists or other personnel.
- *Feature extraction.* This is the fourth step and involves extraction of radiomic features from the ROI, which refer to morphologic and textural characteristics, after which "a study on the correlation among the various features and a first analysis to identify the 'highly' informative features (reliable features referred to as 'radiomic signatures') is applied, and they are selected based on user-defined criteria."
- *Use of machine learning* is the final step, and it is used to "construct, train, and evaluate mathematical models to predict outcomes (e.g., disease classification, treatment response, and survival). At this stage, non-imaging datasets, such as biological and clinical data, may also be integrated to develop holistic radiomics models that can be used as supporting tools for clinical decision-making" [28].

In summary, radiomics is an evolving field of study with the main objective of converting images into mineable data using a series of successive steps via computer processing [29]. These steps are basically image acquisition, ROI identification and segmentation, quantitative image feature extraction, and data mining and informatic analysis [29]. The data can be extracted and used to improve the accuracy of cancer diagnosis, prognosis, and prediction [30].

6.4.1 *Clinical Applications*

The clinical applications of radiomics have been discussed in the literature and focus on cancer management with respect to tumor detection and diagnosis, treatment outcome prediction for decision-making support, risk assessment, and radiogenomics [29]. It is beyond the scope of this chapter to outline the major elements of these applications, and therefore the interested reader should refer to the papers by McCague et al. [28] and by Liu et al. [29] The paper by McCague et al. [28] describes not only oncological applications but also non-oncological applications of radiomics, the latter illustrated in Fig. 6.6. Additionally, readers interested in radiogenomics should refer to a paper by Singh et al. [31]

6.5 Imaging Biobanks

Another application of AI in medical imaging is that of *imaging biobanks*, defined as "organized databases of medical images and associated imaging biomarkers (radiology and beyond) shared among multiple researchers and linked to other

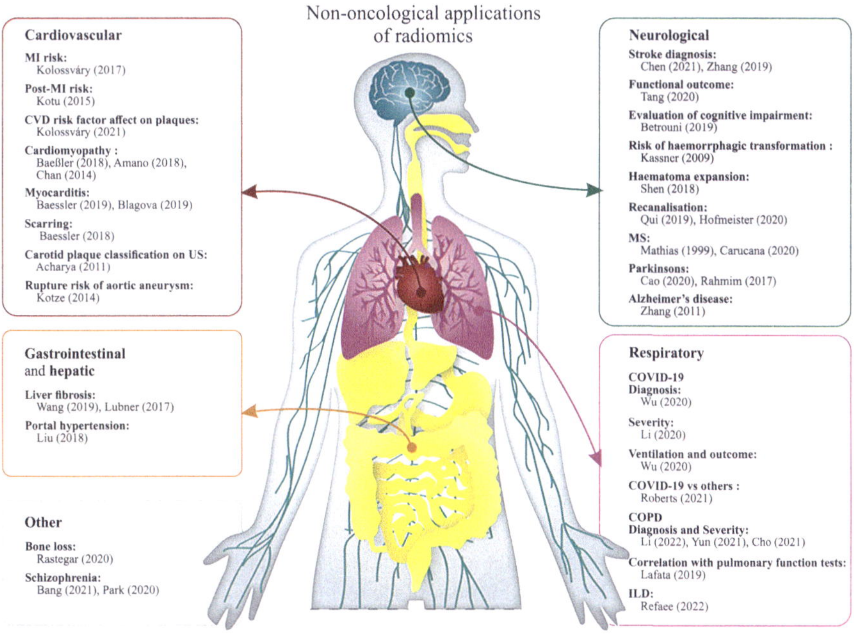

Fig. 6.6 Non-oncological applications of radiomics. (From McCague C, Ramlee S, Reinius M, Selby I, Hulse D, Piyatissa P, Bura V, Crispin-Ortuzar M, Sala E, R. Woitek R (2023). Introduction to radiomics for a clinical audience. Clinical Radiology; 78: (2). 83–98. Reprinted under Creative Commons Attribution license (https://creativecommons.org/licenses/by/4))

biorepositories" [32]. A characteristic feature of imaging biobanks is that the images can be integrated with *biomarkers* and non-imaging data such as clinical and biological data [33].

A biomarker refers to a biological molecule, sometimes called a molecular marker, and it is used to assess how an individual reacts to treatments for a disease or condition. Biomarkers fall into four categories, namely, molecular, physiologic, radiographic, and histologic. While molecular and physiologic examples include blood glucose and blood pressure, respectively, radiographic and histologic examples are bone mineral density and cancer staging, respectively.

The use of biomarkers is now widely accepted in clinical applications and has provided "for better understanding disease processes and the ways in which drugs work to combat disease. This knowledge can be used to diagnose disease earlier or to prevent it before it starts. Biomarkers can be used to improve the efficacy and safety of existing medicines and to develop new medicines. New molecular biomarkers have the potential to personalize disease prevention and management, making the delivery of healthcare more precise, safe, and cost-effective, ultimately improving health outcomes" [34].

In an effort to promote precision medicine for oncologic patients, Borgheresi et al. [35] developed a technological platform that can be used by scientists, clinicians, and others, referred to as NAVIGATOR, the basic framework of which is illustrated in Fig. 6.7, identifying an architecture consisting of a virtual research environment (VRE), an imaging and data biobank, and a patient-centric and data modeling component. This platform offers an open imaging biobank, where patient data can be gathered from multiple sites and shared among multiple related individuals to "experiment, discover, test, and validate novel or known biomarkers over the biobank, via a set of available data analytics tools based on radiomics and AI algorithms" [35].

The kinds of services offered by imaging biobanks are management of image collections, supporting image collections' consultation, relating image cases to external data (data, specimen), and production of imaging biomarkers [36].

6.6 Disease Detection and Classification

The diagnostic process generally consists of two phases, namely, an abductive phase and a deductive phase. While the former addresses the issue of interpreting the information gathered from the patient regarding clinical signs or symptoms, to infer a potential diagnosis, the latter deals with checking assumptions about the diagnosis with specific tests [37]. One such commonplace test is an imaging examination such as diagnostic X-rays; nuclear medicine; diagnostic medical ultrasound; computed tomography (CT); single-photon emission CT (SPECT/CT) and positron emission CT (PET/CT), which are both hybrid imaging modalities; and magnetic resonance imaging (MRI).

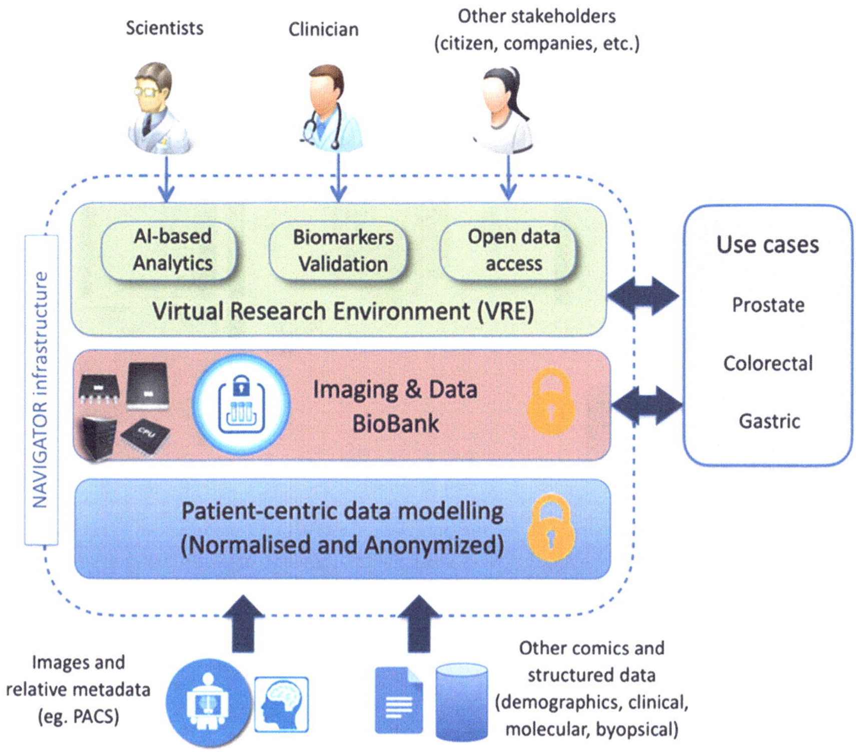

Fig. 6.7 The architectural framework referred to as NAVIGATOR, which offers an open imaging biobank, for use by scientists, clinicians, and other stakeholders. (From Borgheresi, R., Barucci, A., Colantonio, S. et al. NAVIGATOR (2022). An Italian regional imaging biobank to promote precision medicine for oncologic patients. Eur Radio;l Exp 6, 53. https://doi.org/10.1186/s41747-022-00306-9. Reprinted under Creative Commons Attribution License (https://creativecommons.org/licenses/by/4))

The use of AI techniques such as machine learning (ML) and deep learning (DL) have become useful tools in disease detection and classification and have been used for almost 10 years [38, 39], hence providing advantages in the diagnosis of diseases for the effective care and management of the patient [40, 41]. An illustration of using DL for the analysis of COVID-19 using chest X-ray images [42] is shown in Fig. 6.8.

6.6.1 *Framework for AI in Disease Detection Modeling*

A framework for the use of AI in disease detection modeling is illustrated in Fig. 6.9. After the relevant medical datasets (disease symptoms) have been pre-processed (data cleaning and data transformation), they are sent into the AI engine where the

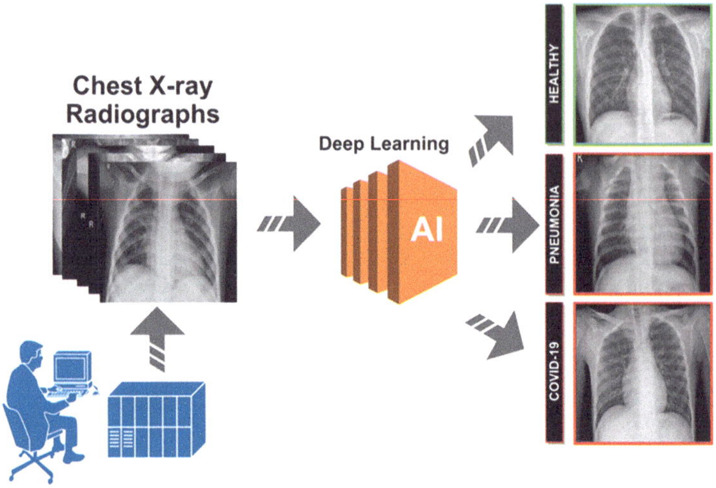

Fig. 6.8 An illustration of using DL for the analysis of COVID-19 using chest X-ray images. (From de Moura J, Novo J, Ortega M (2022) Fully automatic deep convolutional approaches for the analysis of COVID-19 using chest X-ray images. Applied Soft Computing; 115, 108190, https://doi.org/10.1016/j.asoc.2021.108190. Reprinted under Creative Commons Attribution License (https://creativecommons.org/licenses/by/4))

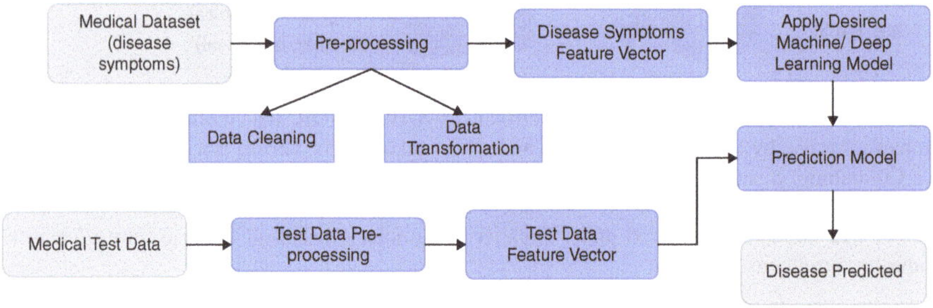

Fig. 6.9 A framework for the use of AI in disease detection modeling. (From Kumar, Y., Koul, A., Singla, R. et al. (2023). Artificial intelligence in disease diagnosis: a systematic literature review, synthesizing framework and future research agenda. Journal of Ambient Intelligence and Humananized Computing; 14:8459–8486. https://doi.org/10.1007/s12652-021-03612-z Reproduced by permission)

appropriate ML or DL model is applied to generate a *predictive model* [43]. Furthermore, pre-processed medical *test data* are input into the predictive model and the outcome is disease prediction. As noted by Toma and Wei [44], predictive modeling is a complex process and uses mathematics and computing to predict future outcomes [44].

Furthermore, as illustrated in Fig. 6.10, through analyses of both historical and current patient clinical data (patient demographics, medical history, and social health factors), predictive models can indicate "patients at higher risk of hospital

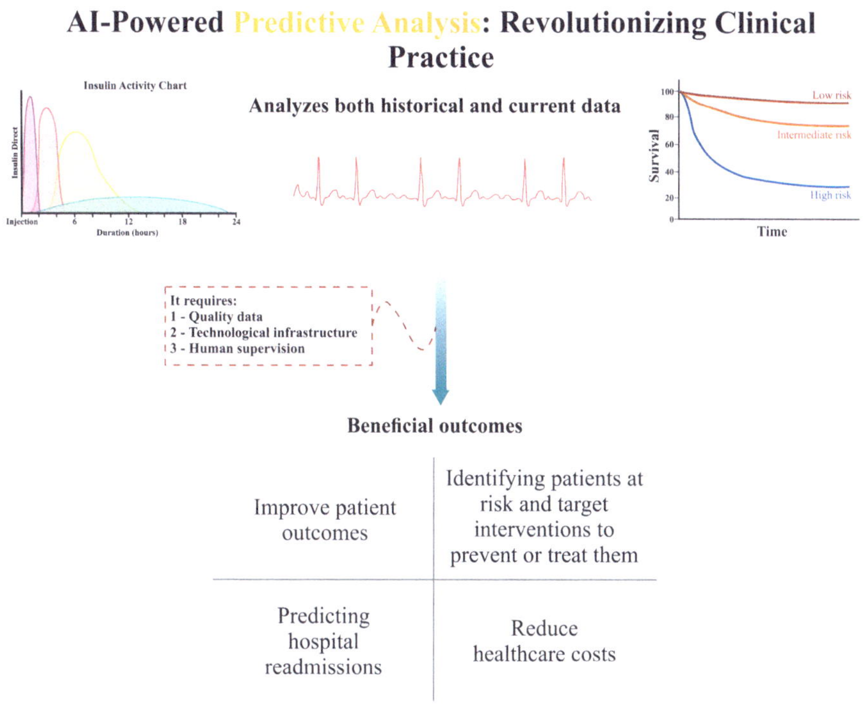

Fig. 6.10 Unlocking the power of patient data with AI-driven predictive analytics. (From Alowais, S.A., Alghamdi, S.S., Alsuhebany, N. et al. Revolutionizing healthcare: the role of artificial intelligence in clinical practice. BMC Med Educ 23, 689 (2023). https://doi.org/10.1186/s12909-023-04698-z Reprinted under Creative Commons Attribution License (https://creativecommons.org/licenses/by/4))

readmissions and target interventions to prevent readmissions," thus cutting the costs of care with better outcomes for the patient [45].

Common ML algorithms used in predictive modeling include support vector machines (SVMs), logistic regression (LR), and clustering4 [46]. These algorithms were briefly described in Chap. 3.

6.6.2 *Evaluation of the Use of AI in Disease Diagnosis*

Studies have been done highlighting the use of AI in disease diagnosis [44–47]. It is not within the scope of this chapter to outline details of such studies; however, the interested reader may refer to a comprehensive review of this topic by Nia et al. [48] This review in particular not only examines the use of AI in disease diagnosis and prediction of cancers, neural disorders, and other diseases but also compares the performance of ML and DL techniques, such as the support vector machine (SVM)

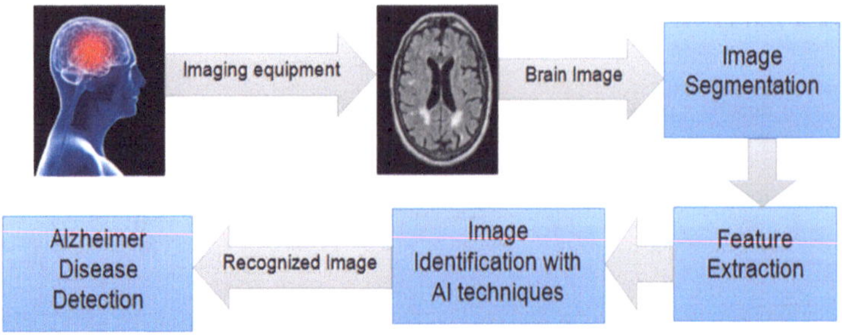

Fig. 6.11 An overview of the use of AI techniques in the detection of Alzheimer's disease. (From Subasi A (2020). Use of artificial intelligence in Alzheimer's disease detection, (Chap. 11), Editor; Barh D, Artificial Intelligence in Precision Health, Academic Press; 257-278, https://doi.org/10.1016/B978-0-12-817133-2.00011-2. Reproduced by permission)

algorithm (ML method described in Chap. 3) and the convolutional neural network (CNN) as a DL method and are popular AI algorithms for analyzing and diagnosing diseases [48]. In terms of performance, the authors found that these AI algorithms "perform more precisely compared to specialists without human error" [48].

An overview of the use of AI techniques in the detection of Alzheimer's disease is illustrated in Fig. 6.11. The results of image segmentation and feature extraction performed on the input images are subsequently sent to the AI engine for image identification [49].

6.7 Dose Optimization

Optimization is a guiding principle of radiation protection framework advocated by the International Commission on Radiological Protection (ICRP) and other radiation protection organizations around the world [50]. This principle ensures that radiation doses are kept as low as reasonably achievable (ALARA), taking into account economic and social factors. The ICRP also uses the term optimization of radiation protection (ORP) synonymous with ALARA. To apply this principle in practice, radiation workers must use the least possible dose for an examination without compromising the image quality needed for diagnosis. ALARA addresses the technical elements of radiation protection, which deals with international and national regulations for dose management, equipment specifications, procedures for minimizing dose to patients and personnel, and quality control considerations [50].

Several studies have reported on the role of AI in dose optimization, not only in CT image reconstruction but also in "clinical workflow optimization, through improvements in patient positioning and image acquisition" [51]. Most studies, however, have focused on the role of AI in CT dose optimization [52] and in fluoroscopy [53–55]. These studies report that DL algorithms (Chap. 4) are the most popular algorithms used in dose optimization, and most studies demonstrated that

AI could reduce radiation dose by 36–70% without losing diagnostic information [52]. The use of AI in CT dose optimization will be described in Chap. 8.

6.7.1 AI in Fluoroscopy

The radiation dose to patients during fluoroscopy has been reported to be extremely high compared with most procedures in radiology, and some patients have experienced radiation-induced skin burns [56]. Additionally, occupational exposures are highest in fluoroscopy [57].

A recent study by Bang et al. [54] described the use of AI to reduce the radiation dose in fluoroscopy-guided endoscopic procedures. They used a system referred to as the *Omega FluoroShield*™ which received FDA clearance in 2020, which is a convolutional neural network (CNN) DL algorithm, for the analysis of fluoroscopy images. The dose reduction to patients was significantly lower compared to conventional fluoroscopy procedures [55].

6.8 Structured Reporting

When patient radiology images are interpreted and reported using radiology diagnostic lexicon, *conventional narrative reports* (free-text reports (FTRs)) are created and filed for retrospective analysis by referring clinicians concerned about the patients' medical condition. These reports contain at least five headings, namely, indication, technique, comparison, findings, and impression. It has been found that these narratives may lead to "excessive variability in the language, length, and style, which can minimize report clarity and make it difficult for referring clinicians to identify key information needed for patient care" [58]. This limitation has been overcome through the development and use of what is referred to as *structured reporting*, another useful tool to the efficiency of clinical and radiological workflow [59].

6.8.1 Definition

Several definitions of the term "structured reporting" [59] are the following:

1. "A report is qualified as structured when all of the relevant information and diagnostic impressions are included, following specific terms and descriptors previously defined K, as well as a predefined design" [60].
2. Structured reporting is "the creation of standardized, organized information from templates via menus into a natural-sounding language report" [61].

3. "Structured reporting means the use of predefined formats and terms to create reports; in this sense, structured reports are those based on templates or checklists" [62].

The above definitions lead to yet another definition proposed by Nobel et al. [59] who state that "structured reporting is the use of an information technology (IT)-based means of importing and arranging medical content in the radiological report."

More recently, Jorg et al. [63] have developed an AI-based tool based on speech recognition and natural language processing, an example of which is illustrated in Fig. 6.12. Essentially, the reporting tool consists of two major components, namely, a frontend and a backend, containing structures designed to facilitate efficient reporting as shown in Fig. 6.13. For more details of this structured reporting AI-based tool, interested readers should refer to the paper by Jorg et al. [63]

6.9 Image Processing

Conventional *image processing* in medical imaging has become commonplace because all images are in digital format. A comprehensive paper of digital image processing is one by Seeram and Seeram [64], who describe the details of five digital *image processing classes* as identified by image processing experts Baxes [65] and Gonzalez and Woods [66]. These include image enhancement, image restoration, image analysis, image compression, and image synthesis, which are used to transform an input image into an output image to suit the needs of the observer. Furthermore, image processing can also extract information from digital images by using image segmentation, feature extraction, and image reconstruction. These techniques not only allow us to process digital images to improve their quality and

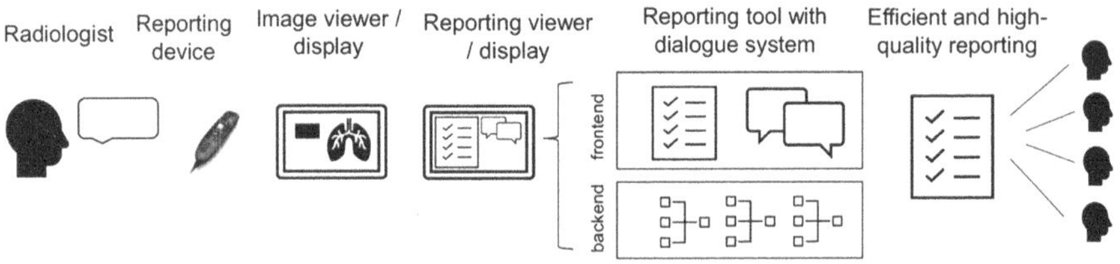

Fig. 6.12 Graphic representation of the reporting workflow using the reporting tool. Reporting is done as free text using speech recognition. The reporting tool comprises a user interface (frontend) and a dialogue system that enables NLP-based conversion of free text into a structured report and communication with the radiologist (backend). The combination of these components enables efficient and high-quality reporting. (From Jorg, T., Kämpgen, B., Feiler, D. et al. Efficient structured reporting in radiology using an intelligent dialogue system based on speech recognition and natural language processing. Insights Imaging 14, 47 (2023). https://doi.org/10.1186/s13244-023-01392-y. Reprinted under Creative Commons Attribution License (https://creativecommons.org/licenses/by/4))

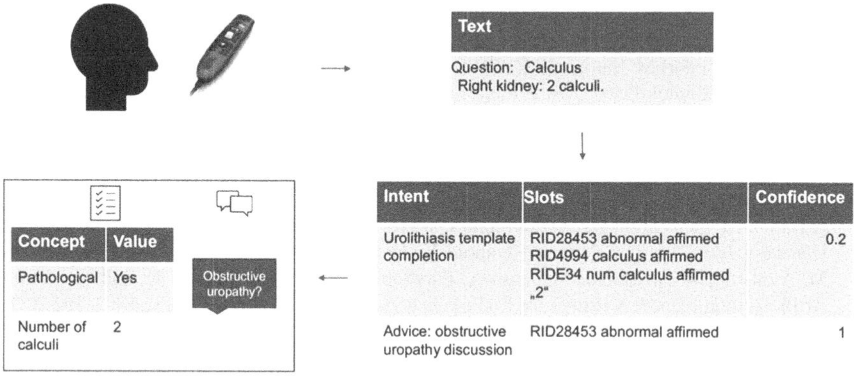

Fig. 6.13 An example of the structured reporting workflow. The user has reported two calculi in the right kidney. Speech is translated to text (speech recognition), and text is translated into structured content of intents and slots (NLP and dialogue management). (From Jorg, T., Kämpgen, B., Feiler, D. et al. Efficient structured reporting in radiology using an intelligent dialogue system based on speech recognition and natural language processing. Insights Imaging 14, 47 (2023). https://doi.org/10.1186/s13244-023-01392-y. Reprinted under Creative Commons Attribution License (https://creativecommons.org/licenses/by/4))

remove noise and artifacts but also to "extract meaningful information and metadata and recognize patterns and objects" [67].

6.9.1 *Major Role of AI in Medical Image Manipulation and Comprehension*

AI is used in medical image manipulation and comprehension, as covered in detail in Chap. 5. In summary, this area of image processing includes pre-processing operations such as those used to reduce noise, and image enhancement, segmentation, feature extraction, and representation within medical images, and in particular medical image classification, and highlights the interpretability and explainability of deep learning models. Finally, medical image segmentation is described focusing on pixel-level and object-level segmentation approaches, popular architectures such as U-Net, and the manifold applications of segmentation in the landscape of medical imaging.

References

1. Najjar R (2023). Redefining Radiology: A Review of Artificial Intelligence Integration in Medical Imaging. Diagnostics; 3, 2760. https://doi.org/10.3390/diagnostics13172760
2. European Society of Radiology (ESR) (2019) What the radiologist should know about artificial intelligence—an ESR white paper. Insights Imaging 10, 44

3. Hosny A (2018) Parmar, C.; Quackenbush, J.; Schwartz, L.H.; Aerts, H.J. Artificial intelligence in radiology. Nat. Rev. Cancer 2018, 18, 500–510.
4. Pesapane F, Codari M, Sardanelli F (2018) Artificial intelligence in medical imaging: threat or opportunity? Radiologists again at the forefront of innovation in medicine. Eur Radiol Exp 2, 35. https://doi.org/10.1186/s41747-018-0061-6
5. Kim M, Yun J, Cho Y, Shin K, Jang R, Bae H-j, Kim N (2019) Deep Learning in Medical Imaging. Neurospine;16(4):657–668. https://doi.org/10.14245/ns.1938396.198
6. Bhatia H, Bhatia A, Ahuja CK, Singh A, Kushaljit S. Sodhi KS (2022). Artificial Intelligence: A Primer for the Radiologist. Indographics; 1:215–221.
7. Haneberg AG, Pierre K, Winter-Reinhold E, Hochhegger B, Peters KR, Grajo J, Arreola M, Asadizanjani N, Bian J, Mancuso A, Forghani R (2023) Introduction to Radiomics and Artificial Intelligence: A Primer for Radiologists, Seminars in Roentgenology; 58: 2, 152–157.
8. The Best Practices for Improving Radiology Workflow (2022) https://www.ramsoft.com/radiology-workflow Accessed January 2024
9. Mese I, Taslicay CA, Sivrioglu AK (2023) Improving radiology workflow using ChatGPT and artificial intelligence. Clinical Imaging; 103, 109993
10. Pierre K, Haneberg AG, Kwak S, Peters KR, Hochhegger B, Sananmuang T, Tunlayadechanont P, Tighe PJ, Mancuso A, Forghani R (2023). Applications of Artificial Intelligence in the Radiology Roundtrip: Process Streamlining, Workflow Optimization, and Beyond. Seminars in Roentgenology; 58: (2), 158–169.
11. Ranschaert E, Topff L, Pianykh O (2021). Optimization of Radiology Workflow with Artificial Intelligence. Radiol Clin North Am; 59 (6):955–966. https://doi.org/10.1016/j.rcl.2021.06.006. PMID: 34689880
12. Li Y, Gao W, Luan Z, Zhou Z, Li J. (2023). The Impact of Chat Generative Pre-trained Transformer (ChatGPT) on Oncology: Application, Expectations, and Future Prospects. Cureus;15(11): e48670. https://doi.org/10.7759/cureus.48670. PMID: 38090410; PMCID: PMC10714025
13. Meşe İ, Taşlıçay CA, Kuzan BN, Kuzan T, Sivrioğlu AK. Educating the next generation of radiologists: a comparative report of ChatGPT and e-learning resources. Diagn Interv Radiol. 25 December 2023 https://doi.org/10.4274/dir.2023.232496
14. Hetler A. (2024) What is ChatGPT? https://www.techtarget.com/whatis/definition/ChatGPT Accessed January 2024
15. Open AI (2024) https://openai.com/about Accessed January 2024
16. Doi K (2007). Computer-aided diagnosis in medical imaging: historical review, current status and future potential. Comput Med Imaging Graph; 31(4–5):198–211.
17. Takahashi R, Kajikawa Y (2017) Computer-aided diagnosis: A survey with bibliometric analysis. International journal of medical informatics. 101: 58–67.
18. Winsberg F, Elkin M, Macy J, et al (1967). Detection of radiographic abnormalities in mammograms by means of optical scanning and computer analysis. Radiology; 89:211–215
19. Gao Y, Geras KJ, Lewin AA, Moy L (2019). New Frontiers: An Update on Computer-Aided Diagnosis for Breast Imaging in the Age of Artificial Intelligence. AJR; 212:300–307
20. Cho HC, Hadjiiski L, Sahiner B, et al (2011). Similarity evaluation in a content-based image retrieval (CBIR) CADx system for characterization of breast masses on ultrasound images. Med Phys; 38:1820–1831
21. Giger ML (2018). Machine learning in medical imaging. J Am Coll Radiol; 15(3 Pt B):512–520
22. Santos MK, Ferreira Júnior JR, Wada DT, Tenório APM, Barbosa MHN, Marques PMA (2019). Artificial intelligence, machine learning, computer-aided diagnosis, and radiomics: advances in imaging towards to precision medicine. Radiol Bras;52(6):387–396. https://doi.org/10.1590/0100-3984.2019.0049. Erratum in: Radiol Bras. 2022 May-Jun;55(3):208. PMID: 32047333; PMCID: PMC7007049.
23. Zhang W, Guo Y, Jin Q (2023). Radiomics and Its Feature Selection: A Review. Symmetry; 15(10):1834. https://doi.org/10.3390/sym15101834
24. Scapicchio, C., Gabelloni, M., Barucci, A. et al (2021). A deep look into radiomics. Radiol med 126, 1296–1311. https://doi.org/10.1007/s11547-021-01389-x

25. Lambin P, Rios-Velazquez E, Leijenaar R, et al (2012). Radiomics: extracting more information from medical images using advanced feature analysis. Eur J Cancer 48(4):441–446.
26. Taged P, Kumar V, Gu Y, Basu S, et al (2012). Radiomics: the process and the challenges. Magn Reson Imaging 30(9):1234–1248, 2012
27. LeCun Y, Bengio Y, Hinton G (2015). Deep learning. Nature 521(7553):436–444.
28. McCague C, Ramlee S, Reinius M, Selby I, Hulse D, Piyatissa P, Bura V, Crispin-Ortuzar M, Sala E, Woitek R (2023). Introduction to radiomics for a clinical audience. Clinical Radiology; 78: (2). 83–98.
29. Liu Z, Wang S, Dong D, Wei J, Fang C, Zhou X, Sun K, Li L, Li B, Wang M, Tian J (2019). The Applications of Radiomics in Precision Diagnosis and Treatment of Oncology: Opportunities and Challenges. Theranostics;9(5):1303–1322. https://doi.org/10.7150/thno.30309. PMID: 30867832; PMCID: PMC6401507.
30. Huang Hong H (2022). Editorial: The Application of Radiomics and Artificial Intelligence in Cancer Imaging. Frontiers in Oncology; 12 URL=https://www.frontiersin.org/articles/10.3389/fonc.2022.864940 https://doi.org/10.3389/fonc.2022.864940
31. Singh A, Chitalia R, Kontos D (2021). Radiogenomics in brain, breast, and lung cancer: opportunities and challenges. J Med Imaging (Bellingham);8(3):031907. https://doi.org/10.1117/1.JMI.8.3.031907. Epub 2021 Jun 18. PMID: 34164563; PMCID: PMC8212946.
32. Mission Statement of the ESR Working Group on Imaging Biobanks (2014) https://www.myesr.org/cms/website.php?id=/en/research/esr_wg_on_imaging_biobanks/esr.htm
33. Gabelloni M, Faggioni L, Borgheresi R, Restante G, Shortrede J, Tumminello L, Scapicchio C, Coppola F, Cioni D, Gómez-Rico I, Martí-Bonmatí L, Neri E. Bridging gaps between images and data: a systematic update on imaging biobanks. Eur Radiol. 2022 May;32(5):3173–3186. https://doi.org/10.1007/s00330-021-08431-6. Epub 2022 Jan 10. PMID: 35001159.
34. Battineni G, Hossain MA, Chintalapudi N, Amenta F. A Survey on the Role of Artificial Intelligence in Biobanking Studies: A Systematic Review. Diagnostics (Basel). 2022 May 9;12(5):1179. https://doi.org/10.3390/diagnostics12051179. PMID: 35626333; PMCID: PMC9140088.
35. Haskel R (2019) What is a Biomarker? Lippincott's Nursing Center. https://www.nursingcenter.com/blog Accessed January 24, 2024
36. European Society of Radiology (ESR). ESR Position Paper on Imaging Biobanks. Insights Imaging 6, 403–410 (2015). https://doi.org/10.1007/s13244-015-0409-x
37. Baerheim A. The diagnostic process in general practice: has it a two-phase structure? Family Practice 2001; 18: 243–245.
38. Nasser A, Akhloufi MA (2023) A Review of Recent Advances in Deep Learning Models for Chest Disease Detection Using Radiography. Diagnostics; 13(1):159. https://doi.org/10.3390/diagnostics13010159
39. Kabiraj, A., Meena, T., Reddy, P.B., Roy, S. (2022). Detection and Classification of Lung Disease Using Deep Learning Architecture from X-ray Images. In: Bebis, G., et al. Advances in Visual Computing. ISVC 2022. Lecture Notes in Computer Science, vol 13598. Springer, Cham. https://doi.org/10.1007/978-3-031-20713-6_34
40. Mirbabaie M, Stieglitz S, Nicholas R. J. Frick NRJ (2021) Artificial intelligence in disease diagnostics: A critical review and classification on the current state of research guiding future direction Health and Technology; 11:693–731; https://doi.org/10.1007/s12553-021-00555-5
41. Okeibunor JC, Jaca A, Iwu-Jaja CJ, Idemili-Aronu N, Ba H, Zantsi ZP, Ndlambe AM, Mavundza E, Muneene D, Wiysonge CS and Makubalo L (2023) The use of artificial intelligence for delivery of essential health services across WHO regions: a scoping review. Front. Public Health 11:1102185. https://doi.org/10.3389/fpubh.2023.1102185
42. de Moura J, Novo J, Ortega M (2022) Fully automatic deep convolutional approaches for the analysis of COVID-19 using chest X-ray images. Applied Soft Computing; 115, 108190, https://doi.org/10.1016/j.asoc.2021.108190
43. Kumar, Y., Koul, A., Singla, R. et al (2023). Artificial intelligence in disease diagnosis: a systematic literature review, synthesizing framework and future research agenda. Journal of

Ambient Intelligence and Humananized Computing; 14:8459–8486 https://doi.org/10.1007/s12652-021-03612-z
44. Toma M, Wei OC. Predictive Modeling in Medicine. Encyclopedia. 2023; 3(2):590–601. https://doi.org/10.3390/encyclopedia3020042
45. Alowais, S.A., Alghamdi, S.S., Alsuhebany, N. et al. Revolutionizing healthcare: the role of artificial intelligence in clinical practice. BMC Med Educ 23, 689 (2023). https://doi.org/10.1186/s12909-023-04698-z
46. Battineni G, Sagaro GG, Chinatalapudi N, Amenta F (2020). Applications of Machine Learning Predictive Models in the Chronic Disease Diagnosis. Journal of Personalized Medicine. 2020; 10(2):21. https://doi.org/10.3390/jpm10020021
47. Muddana, A.L., Chennam, K.K., Revathi, V. (2021). Artificial Intelligence for Disease Identification and Diagnosis. In: Siarry, P., Jabbar, M., Aluvalu, R., Abraham, A., Madureira, A. (eds) The Fusion of Internet of Things, Artificial Intelligence, and Cloud Computing in Health Care. Internet of Things. Springer, Cham. https://doi.org/10.1007/978-3-030-75220-0_9
48. Nia NG, Kaplanoglu E, Nasab A (2023). Evaluation of artificial intelligence techniques in disease diagnosis and prediction. Discov Artif Intell; 3, 5. https://doi.org/10.1007/s44163-023-00049-5
49. Subasi A (2020). Use of artificial intelligence in Alzheimer's disease detection, (Chapter 11), Editor; Barh D, Artificial Intelligence in Precision Health, Academic Press; 257–278, https://doi.org/10.1016/B978-0-12-817133-2.00011-2
50. Seeram E (2020) Rad Tech's Guide to Radiation Protection. Hoboken, NJ. John Wiley & Sons Ltd.
51. Caruso, D., De Santis, D., Polidori, T., Zerunian, M., Laghi, A. (2022). Radiation Dose Optimization: The Role of Artificial Intelligence. In: De Cecco, C.N., van Assen, M., Leiner, T. (eds) Artificial Intelligence in Cardiothoracic Imaging. Contemporary Medical Imaging. Humana, Cham. https://doi.org/10.1007/978-3-030-92087-6_17
52. Ng CKC (2022) Artificial Intelligence for Radiation Dose Optimization in Pediatric Radiology: A Systematic Review. Children; 9(7):1044. https://doi.org/10.3390/children9071044
53. Shim JH, Choi SY, Chang IH, Park SB (2023) Dose Optimization Using a Deep Learning Tool in Various CT Protocols for Urolithiasis: A Physical Human Phantom Study. Medicina; 59(9):1677. https://doi.org/10.3390/medicina59091677
54. Bang JY, Hough M, Hawes R, Varadarajulu S (2020) Use of Artificial Intelligence to Reduce Radiation Exposure at Fluoroscopy-Guided Endoscopic Procedures. The American Journal of Gastroenterology 115(4):p 555–561, https://doi.org/10.14309/ajg.0000000000000565
55. Fiagbedzi, E., Gorleku, P., Nyarko, S., Atuwo-Ampoh, V., Fiagan, Y. and Asare, A. (2022) The Role of Artificial Intelligence (AI) in Radiation Protection of Computed Tomography and Fluoroscopy: A Review. Open Journal of Medical Imaging, 12, 25–36. https://doi.org/10.4236/ojmi.2022.121004.
56. Bushberg JT, Seibert AJ, Leidholdt Jr. EM, Boone JM. The Essential Physics of Medical Imaging. Philadelphia, PA, Wolters Kluwerl Lippincott Williams & Wilkins. 2020
57. Bushong S. Radiologic Science for Technologists, 12th Edition. Elsevier: St Louis, MO. 2023
58. Ganeshan D, Duong PT, Probyn L, Lenchik L, McArthur TA, Retrouvey M, Ghobadi EH, Desouches SL, Pastel D, Francis IR (2018). Structured Reporting in Radiology. Acad Radiol;25(1):66–73. https://doi.org/10.1016/j.acra.2017.08.005. Epub 2017 Oct 10. PMID: 29030284
59. Nobel JM, van Geel K, Robben SGF (2022) Structured reporting in radiology: a systematic review to explore its potential. Eur Radiol;32(4):2837–2854. https://doi.org/10.1007/s00330-021-08327-5. Epub 2021 Oct 15. PMID: 34652520; PMCID: PMC8921035
60. Nobel JM, Kok EM, Robben SGF (2020) Redefining the structure of structured reporting in radiology. Insights Imaging 11:10
61. Moher D, Liberati A, Tetzlaff J, Altman DG; PRISMA Group (2009) Preferred reporting items for systematic reviews and meta-analyses: the PRISMA statement. BMJ; 339:b2535
62. Mays N, Pope C, Popay J (2005) Systematically reviewing qualitative and quantitative evidence to inform management and policy-making in the health field. J Health Serv Res Policy 10(Suppl 1):6–20

63. Jorg T, Kämpgen B, Feiler D. et al. (2023). Efficient structured reporting in radiology using an intelligent dialogue system based on speech recognition and natural language processing. Insights Imaging;14, 47. https://doi.org/10.1186/s13244-023-01392-y
64. Seeram E, Seeram D (2008). Image Postprocessing in Digital Radiology: A Primer for Technologists. Journal of Medical Imaging and Radiation Sciences; 39: 23–41
65. Baxes, G. A. (1994). Digital image processing: principles and applications. New York: John Wiley & Sons
66. Gonzalez, R. C., & Woods, R. E. (2008). Digital image processing. Harlow: Prentice-Hall
67. Vegavid (2024). The Power of AI in Image Processing: A Comprehensive Guide https://vegavid.com/blog/power-of-ai-in-image-processing/# Accessed Feb 2024

Chapter 7
Artificial Intelligence in Computed Tomography Image Reconstruction

7.1 Introduction

A historical timeline of the key discoveries in medical imaging [1] is illustrated in Fig. 6.1. Two important points to note in this time line is the invention of the CT scanner in 1971 and the approval of artificial intelligence (AI) algorithms for medical imaging, in 2018, by the Food and Drug Administration (FDA) in the United States.

At the time of invention, the CT scanner was called the EMI scanner, since it was developed at the Electrical and Music Industry (EMI) Central Research Laboratories in England, by Godfrey Newbold Hounsfield who made the scanner clinically useful. For this invention, Hounsfield shared the Nobel Prize for Physiology or Medicine in 1979, with Allan Cormack, a medical physicist in South Africa. Both Nobel Laureates worked on the mathematics of image reconstruction, a major principle driving the efficacy of the CT scanner [2, 3].

The increased use of the CT scanner in clinical medicine sparked concerns about high radiation doses to patients from CT scanning, compared with other X-ray imaging modalities. With this in mind, a number of significant technical innovations were introduced to optimize patient doses [4, 5]. One such innovation is image reconstruction used in CT, the overall goal of which is to produce high-quality images in low-dose CT (LDCT) imaging.

The purpose of this chapter is to describe CT image reconstruction techniques in CT, from the early filtered back projection and recent iterative reconstruction algorithms to the current state-of-the-art algorithms based on AI principles.

E. Seeram, V. Kanade, *Artificial Intelligence in Medical Imaging Technology*,
https://doi.org/10.1007/978-3-031-64049-0_7

7.2 The Major Components of the CT Scanner

An overview of the major technical components of the CT scanner is shown in Fig. 7.1. These include data acquisition, image reconstruction, and image display, storage, and communication While data acquisition system collects attenuation data (measured projection profiles) from the patient during scanning, image reconstruction is based on mathematics that use the projection profiles to create CT images. These images are subsequently displayed for viewing and interpretation.

Image reconstruction is the heart of the CT scanner, since images can only be produced using these algorithms, as illustrated in Fig. 7.2. Without image reconstruction, a sinogram is obtained using the four measured projection profiles. When the sinogram is used as input into the image reconstruction algorithm, a CT image is obtained.

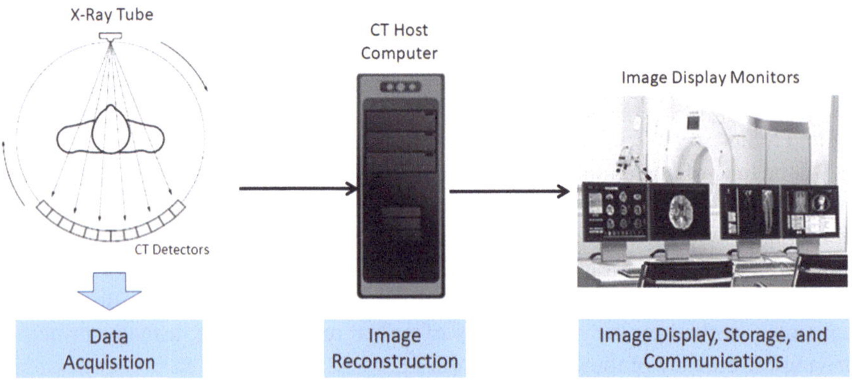

Fig. 7.1 The major system components of a CT scanner, namely, data acquisition components, image reconstruction algorithm, and image display components that allow image viewing and interpretation

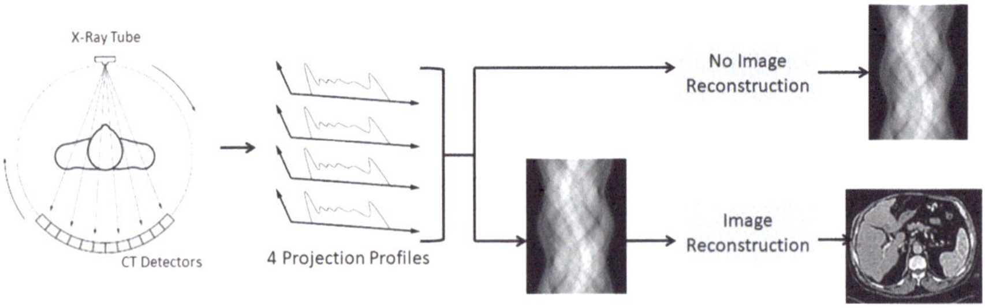

Fig. 7.2 The use of measured projection profiles with image reconstruction produces a diagnostic CT image

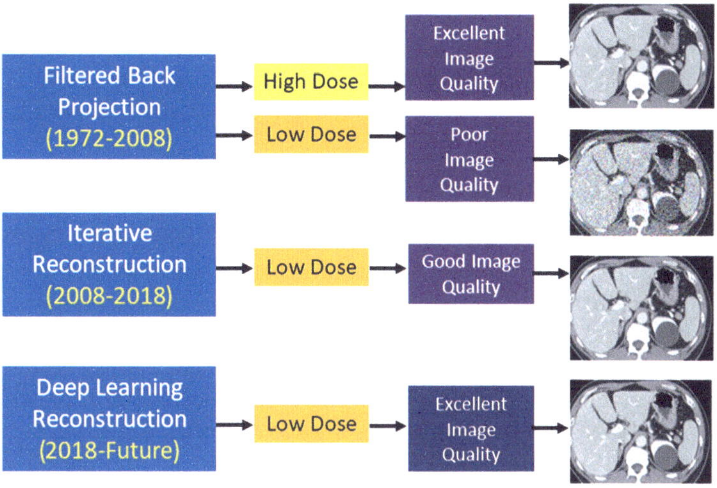

Fig. 7.3 Three periods of CT image reconstruction algorithms used in clinical imaging. See text for further explanation

7.3 CT Image Reconstruction Fundamentals

Image reconstruction in CT is based on mathematics and is defined as "the process of producing an image in a two-dimensional distribution (usually of some physical property), from estimates of its line integrals along a finite number of lines of known locations" [6]. In CT, the physical property is the linear attenuation coefficient (μ) of the tissues; the line integrals refer to the sum of the attenuation along each ray in the beam that passes through the slice; and the finite number of lines refer to the data collected at known positions of the X-ray tube and detectors as they rotate around the patient.

Image reconstruction algorithms (an algorithm is a finite set of rules for solving a problem) provide the solution to the mathematical problem in CT. These algorithms have evolved through the years with the goal of using low doses to produce excellent diagnostic image quality as illustrated in Fig. 7.3 which shows three periods of CT image reconstruction, each using a different image reconstruction algorithm. While the first period used the filtered back projection (FBP) algorithm between 1972 and 2008, the second period used iterative reconstruction (IR) algorithms between 2008 and 2018. Finally, the third period falls between 2018 and extends into the future using AI-based image reconstruction algorithms. Furthermore, Fig. 7.3 shows that the FBP algorithm produces excellent image quality at high doses. This raised concerns during this period because of potential biological risks to patients. The use of low doses with the FBP algorithm produced poor image quality. Subsequently, other algorithms such as IR algorithms were introduced, and as noted in Fig. 7.3, low doses coupled with IR algorithms produced good image quality. Finally, AI-based image reconstruction algorithms (deep learning image reconstruction algorithms) were introduced to address image quality problems created by IR algorithms. Figure 7.3 illustrates that at low doses, using AI-based image reconstruction algorithms produced excellent image quality [7–9].

7.3.1 The Filtered Back Projection Algorithm

The FBP image reconstruction has its roots in an earlier algorithm referred to as the back projection (BP) algorithm and has been the standard CT image reconstruction algorithm for four decades [10, 11]. The major problem with the BP algorithm was that of image blurring that rendered the image useless. The FBP algorithm solves this problem, as illustrated in Fig. 7.4, using four projection profiles, which are subsequently filtered using a digital filter known as *convolution*. The filtered projections are summed, and the negative and positive components are cancelled, hence producing an image free of blurring. Applying a convolutional filter to each set of projection data before back projection is intended to improve sharpness and contrast [12].

The FBP algorithm has limitations. For example, convolutional filters introduce noise and artifacts into the image, especially with LDCT scanning. Other limitations are based on the underlying assumptions about scanner geometry, such as the X-ray focal spot that is a point source, failure to consider shape and size of detector cells and voxels, neglecting image noise resulting from Poisson statistical variations of X-ray photons, and the use of pencil X-ray beam geometry [7, 8, 12]. A significant problem with the FBP algorithm is that in LDCT imaging, using newer advanced CT scanners with improved technology such as faster data collection speeds, is that of increased image noise and artifacts. To solve this problem, IR algorithms were introduced into clinical CT scanning [10, 11, 13].

7.3.2 Iterative Reconstruction Algorithm

The term "iteration" refers to a calculation process using a series of operations repeated several times. IR algorithms use better mathematical estimates and multiple iterations to reduce image noise (Fig. 7.5) while maintaining spatial resolution and image contrast, especially at lower tube currents in LDCT examinations [5, 14, 15]. One notable difference is that instead of filtration using digital filters as in FBP, IR algorithms use better-defined mathematical estimates to provide accurate modeling of the CT system. Furthermore, IR algorithms reduce artifacts caused by metallic implants as well as those resulting from photon starvation and beam-hardening effecs [15].

IR algorithms are currently placed into three categories, namely, statistical methods without modeling, statistical methods with modeling (also referred to as hybrid iterative reconstruction), and pure model-based methods that exceed statistical

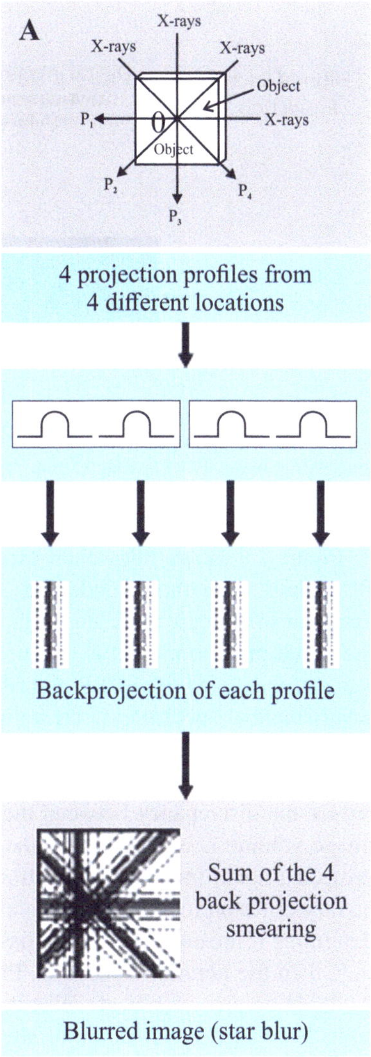

Fig. 7.4 The procedural steps characterized by the FBP algorithm. See text for further explanation

methods. Science defines a model as a representation of an idea, an object, or even a process or a system that is used to describe and explain phenomena that cannot be experienced directly [16]. IR algorithms from four CT vendors are adaptive iterative dose reduction (AIDR) from Canon Medical Systems, advanced modeled iterative reconstruction (ADMIR) from Siemens Healthineers, model-based iterative reconstruction (MBIR) from General Electric HealthCare, and iDose [iterative processing in projection (sinogram) and image domains] from Philips Medical Systems [3]. It is not within the scope of this chapter to describe system modeling, and therefore the interested reader may refer to an article by Hsieh et al. [8]

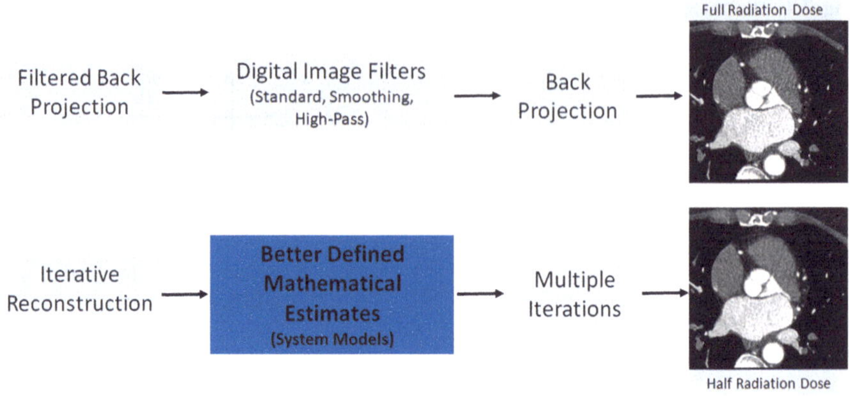

Fig. 7.5 IR algorithms use better mathematical estimates and multiple iterations to reduce image noise while maintaining spatial resolution and image contrast, especially at lower tube currents in LDCT examinations. (Images used with permission of GE HealthCare Technologies Inc.)

Figure 7.6 shows a flowchart explaining the fundamental steps typically used by IR algorithms without modeling. There at least five noteworthy points that are essential for a reasonable understanding of how this algorithm works: Firstly, the measured projection data is the input data, and an initial estimate of the image is reconstructed using the FBP algorithm. Secondly, this image is forward projected (mathematical operation) to create what is referred to as artificial raw data. Thirdly, it involves a comparison of the artificial raw data with the measured projection data, and the difference is referred to as the error or bias. As noted by Hsieh et al. [8], "to reduce the discrepancy between the image and the acquired data, the intermediate image volume is updated using the error. The new intermediate image is forward projected again in the next iteration" [8]. Fourthly, using a predetermined image quality criterion (contained in the algorithm), the CT user makes an assessment to determine if the quality criterion has been met. Fifthly, if this criterion has not been met, then the iteration continues. Finally, if the image quality criterion is not met, the iteration process repeats several times in an iterative cycle until the difference is considered sufficiently minimal, thus terminating the iterative cycle.

IR algorithms perform adequately in LDCT examinations; however, the noise texture has been reported as appearing blotchy, plastic-looking, or unnatural. Furthermore, the amount of degradation is often linked to the strength of the iterative reconstruction algorithm. For example, the stronger the iterative reconstruction, the more plastic the image looks. For this reason, radiologists often limit the iterative reconstruction to a level with which they are comfortable. This creates a trade-off between true dose reduction and reading diagnostic confidence [9, 10, 11, 17]. Additionally, the literature generally shows a consensus that iterative reconstruction algorithms can reduce radiation dose (by 23–76%) and improve image quality in CT in comparison with the FBP algorithm [8, 10, 15, 17, 18].

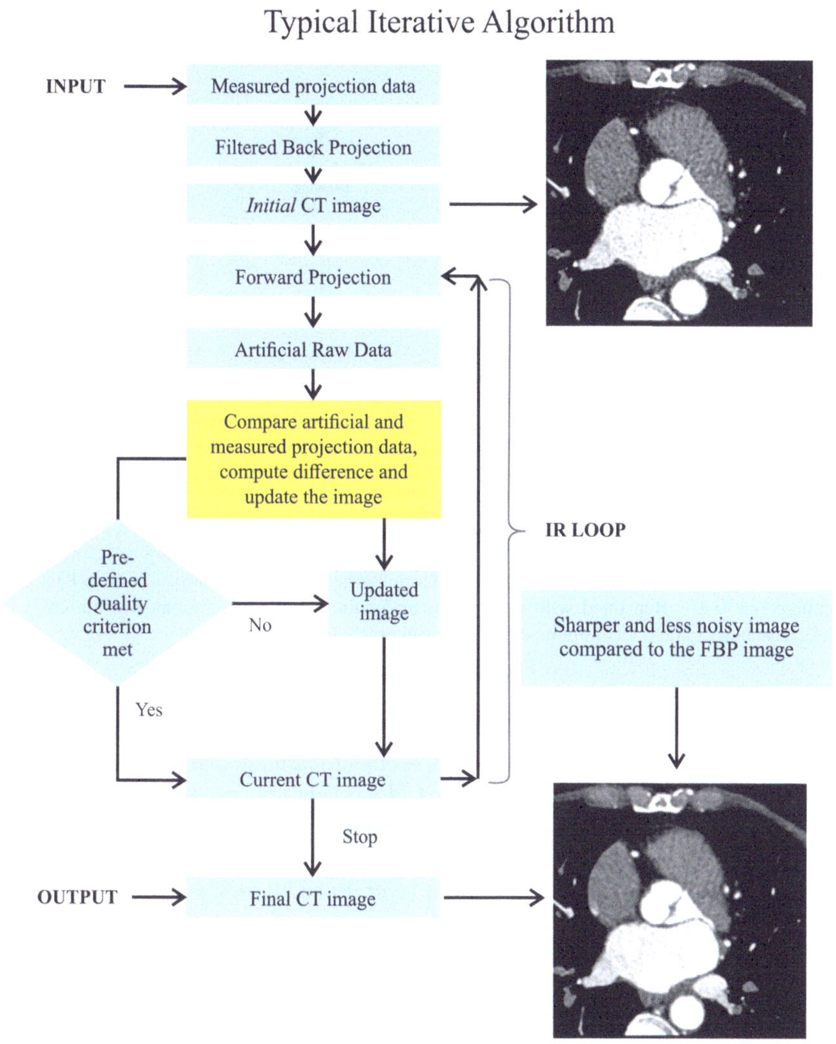

Fig. 7.6 Flowchart illustrating how a typical IR algorithm without modeling works. (Images used with permission of GE HealthCare Technologies Inc.)

7.4 Deep Learning Neural Networks: A Bare Bones Review

The motivation for the application of AI techniques in CT image reconstruction stems from the shortcomings of FBP and IR algorithms mentioned above. The application of AI in CT image reconstruction provides a solution to these shortcomings [8].

AI and its subfields, machine learning (ML) and deep learning (DL), were first introduced in Chaps. 1 and 2, and subsequently ML and DL neural networks were described in detail in Chaps. 3 and 4, respectively. In review, three points are noteworthy: application areas of DL neural networks, use of the convolutional neural

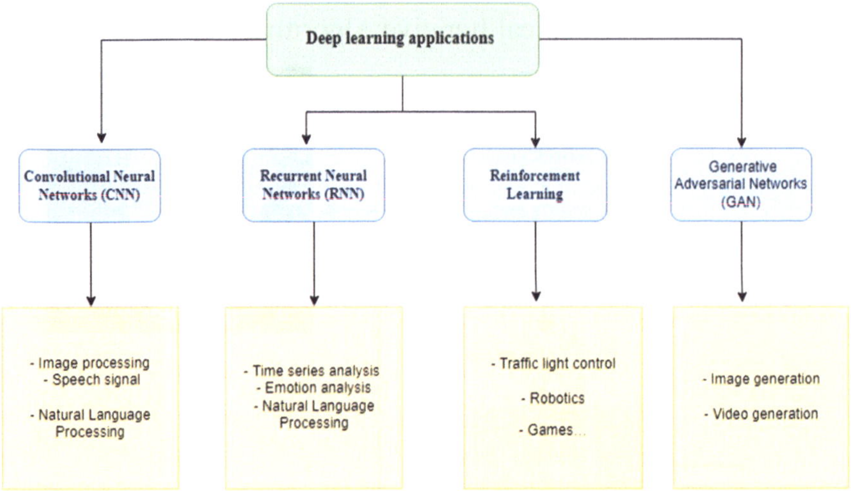

Fig. 7.7 Deep learning application areas. See text for further highlights. (From Taye, Mohammad Mustafa. 2023. "Understanding of Machine Learning with Deep Learning: Architectures, Workflow, Applications and Future Directions" Computers 12, no. 5: 91. https://doi.org/10.3390/computers12050091. Reprinted with permission under the Creative Commons Attribution (CC BY) license https://creativecommons.org/licenses/by/4.0/)

network (CNN), and the steps used by a neural network. These chapters identified a number of application areas (Fig. 7.7) of DL neural networks, including CNNs, recurrent neutral networks (RNNs), reinforcement learning, and generative adversarial networks (GANs). Furthermore, the CNN is generally used for image data for image classification and segmentation. The major components of a CNN are illustrated in Fig. 7.8 and include a convolutional layer, pooling layer, activation function, fully connected layer, and the output layer. In addition, it is important to note the finite number of steps involved in the operation of a neural network, before examining the application of a neural network in CT image reconstruction. These steps are shown in Fig. 7.9.

7.5 Deep Learning Image Reconstruction Algorithms

DL has gained widespread attention in CT, because it is now applied to image reconstruction principles, with the ultimate goal of providing improved image quality, dose performance, and reconstruction speed compared with iterative reconstruction methods in LDCT imaging [1, 8–10, 17, 19, 20]. A visual comparison of the image quality offered by three CT image reconstruction approaches is shown in Fig. 7.10, highlighting the noise reduction and the improvement of sharpness offered by DL image reconstruction algorithm. Figure 7.11 provides another example of the visual comparison of the denoising (more homogeneous liver texture) with the DL image reconstruction algorithm compared with the IR algorithm [13].

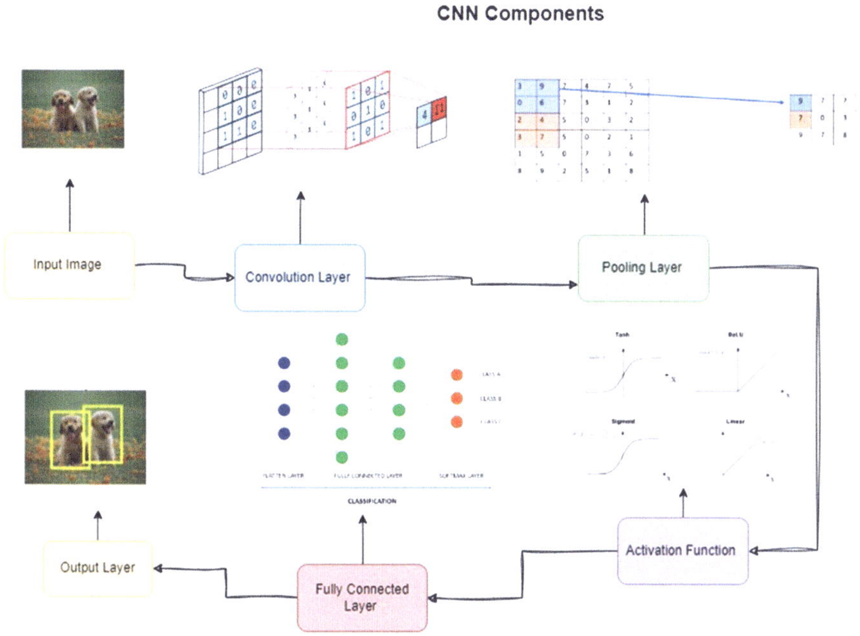

Fig. 7.8 Major components of a CNN. (From Taye, Mohammad Mustafa. 2023. "Understanding of Machine Learning with Deep Learning: Architectures, Workflow, Applications and Future Directions" Computers 12, no. 5: 91. https://doi.org/10.3390/computers12050091. Reprinted with permission under the Creative Commons Attribution (CC BY) license https://creativecommons.org/licenses/by/4.0/)

Furthermore, with both hardware, such as the use of the graphics processing unit (GPU) [9, 13], and software in the reconstruction architecture, as well as the availability of data for training, validation, and testing, DL reconstruction techniques have emerged as practical options for use in CT image reconstruction.

7.5.1 *Basic Concept*

As mentioned earlier, the principles of DL are described in detail in Chap. 4, which described neural networks such as convolutional neural networks (CNNs) and other related topics that are essential for a good understanding how DL can produce images with improved image quality (compared to FBP and IR algorithms) at reduced radiation doses. The basic concept of DL image reconstruction is illustrated in Fig. 7.12, which shows the following: an untrained DL reconstruction engine which contains the neural network, high- and low-quality CT images, a CT scanner that produces measured projection data (sinogram) with low doses, and a trained DL reconstruction engine that outputs the final CT image with improved image quality at low-dose data acquisition. The following steps are essential [13]:

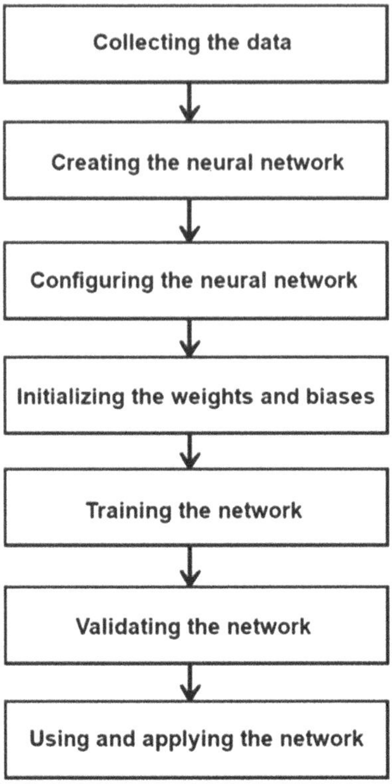

Fig. 7.9 Essential steps of how a neural network operates. (From Kufel J, Bargieł-Łączek K, Kocot S, Koźlik M, Bartnikowska W, Janik M, Czogalik Ł, Dudek P, Magiera M, Lis A, Paszkiewicz I, Nawrat Z, Cebula M, Gruszczyńska K. What Is Machine Learning, Artificial Neural Networks and Deep Learning?—Examples of Practical Applications in Medicine. Diagnostics (Basel). 2023 Aug 3;13(15):2582. https://doi.org/10.3390/diagnostics13152582. PMID: 37568945; PMCID: PMC10417718. (Licensee MDPI, Basel, Switzerland. Reprinted with permission under the Creative Commons Attribution (CC BY) license (https://creativecommons.org/licenses/by/4.0/))

1. Design the neural network architecture, which must consider parameters such as hyperparameters (Chap. 4) which are vital to the function of the network.
2. Train the network using low-quality and high-quality images. The high-quality images are referred to as ground truth images. Ground truth images are considered reference images that have been decided upon by radiologists.
3. After processing, the DL engine produces an output image.
4. This output image is compared with the ground truth image, "calculating the error function to the output" [13].
5. In a process referred to as backpropagation, "the impact of each weight to the error will be calculated and subsequently weights in the network will be adjusted accordingly. This process is repeated iteratively, and the network will successively learn to eliminate most of the noise while keeping anatomical detail" [13].

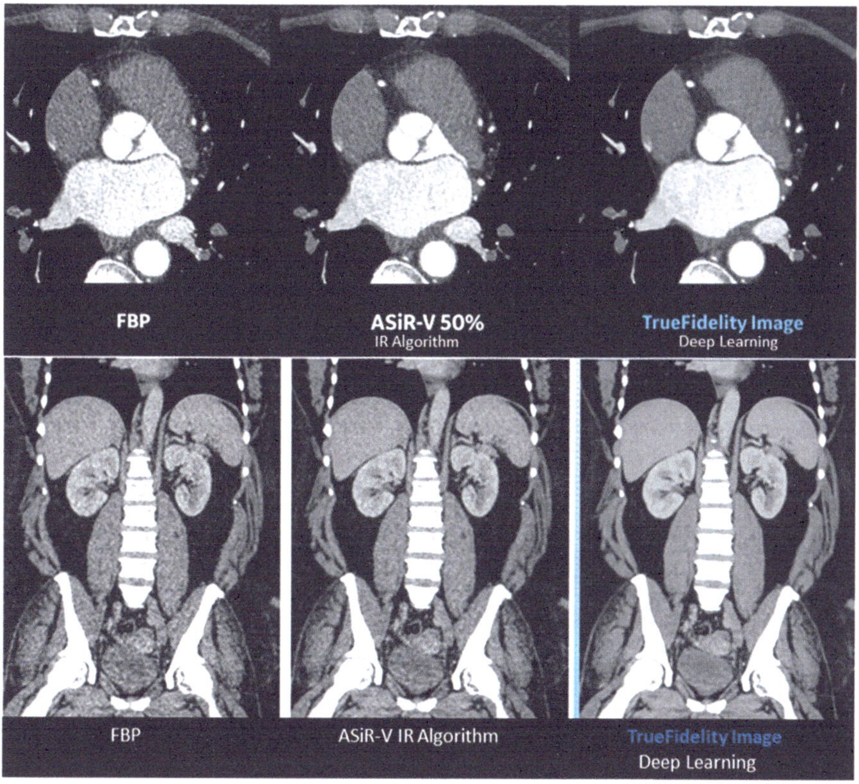

Fig. 7.10 A visual comparison of the image quality offered by three CT image reconstruction approaches (FBP, IR, and DL image reconstruction algorithms) highlighting the noise reduction and the improvement of sharpness offered by DL image reconstruction algorithm. (Images used with permission of GE HealthCare Technologies Inc.)

7.5.2 CT Scanners Using DL Image Reconstruction Algorithms

DL image reconstruction has been implemented by multiple CT vendors and third-party software providers [8, 21, 22]. In 2019, the US Food and Drug Administration (FDA) approved two CT commercial CT scanners using DL algorithms for image reconstruction. These include the Advanced Intelligent Clear-IQ Engine (AiCE) from Canon Medical Systems [21] and TrueFidelity from General Electric (GE) Healthcare [8, 21].

A generalized overview of how both of these two scanners work using DL image reconstruction is illustrated in Fig. 7.13. The overall goal of the DL algorithm in both scanners is to reconstruct images using both low-dose CT data and high-dose raw data (sinogram) to generate what is referred to as the ground truth image. The process is as follows:

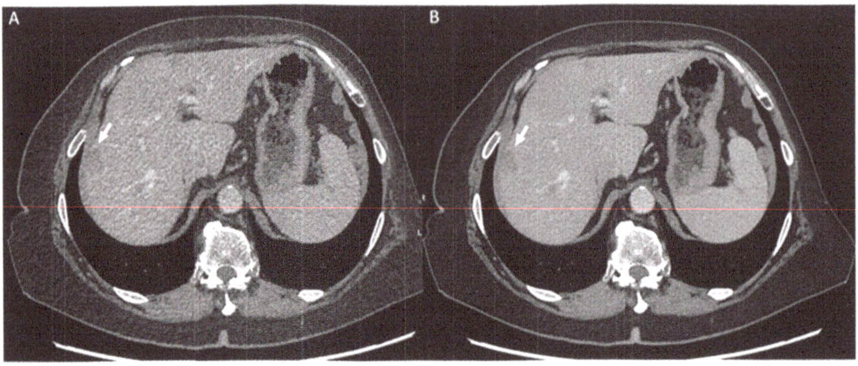

Fig. 7.11 Example of the visual comparison of the denoising (more homogeneous liver texture) with the DL image reconstruction algorithm compared with the IR algorithm. (From Arndt C, Güttler F, Heinrich A etal (2021) Deep Learning CT Image Reconstruction in Clinical Practice. Fortschr Röntgenstr; 193: 252–261. Reproduced by permission)

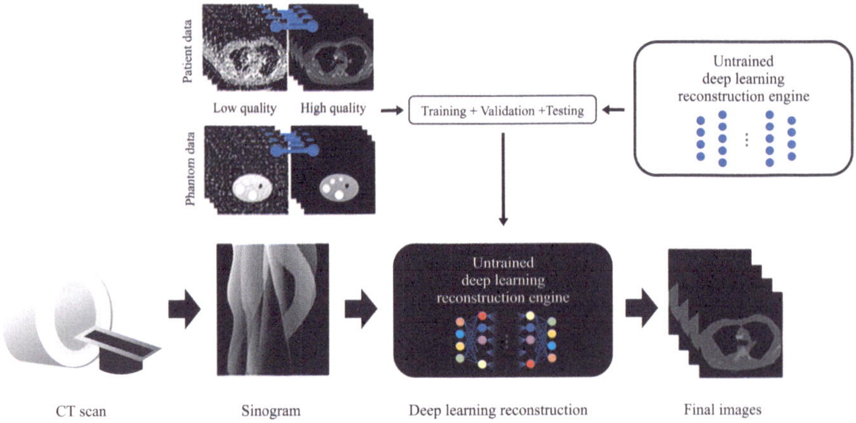

Fig. 7.12 The essential steps of a deep learning image reconstruction. (From Arndt C, Güttler F, Heinrich A etal (2021) Deep Learning CT Image Reconstruction in Clinical Practice. Fortschr Röntgenstr; 193: 252–261. Reproduced by permission)

1. Low-dose raw data (sinogram) are sent to the DL neural network (DLNN).
2. High-dose CT images (ground truth training data) are sent to the DLNN. The high-dose raw data sinogram is reconstructed to produce what is referred to as the ground truth image (reference image) that has been determined by radiologists as the acceptable image required.
3. The DLNN outputs an image referred to as the output DL CT image.
4. The output DL CT image is then compared with the reference image and the amount of error is noted.
5. The process repeats itself (via back propagation) until the output DL image matches the reference image.

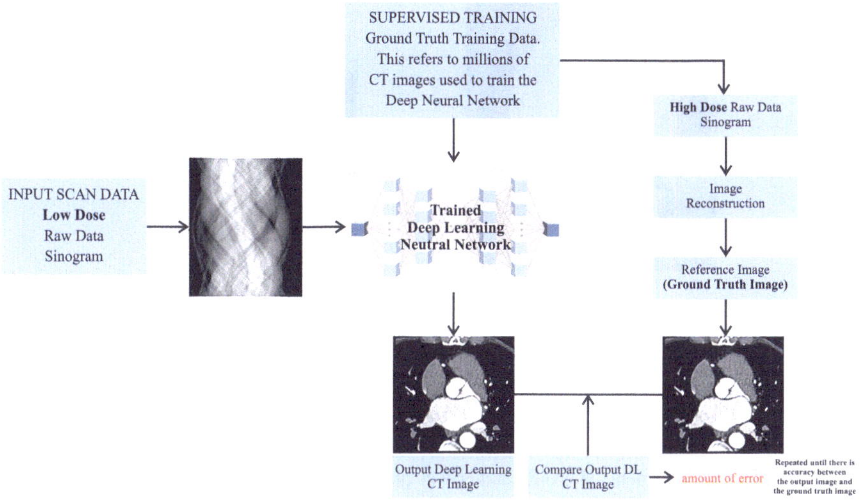

Fig. 7.13 A generalized overview the steps involved in the training and optimization of the DRL algorithm in CT image reconstruction. (Images used with permission of GE HealthCare Technologies Inc.)

During the training process, the output images are compared with a reference image using various parameters, including image noise, noise texture, low contrast resolution, and low contrast detectability. Differences are reported by the output image to the network via a procedure referred to as backpropagation that strengthens some equations and weakens others and tries again. This process repeats itself until the output image is an accurate representation of the ground truth (true values) image. During the final step, performance or verification, the algorithm is required to reconstruct clinical and phantom cases it has never encountered, including rare cases.

The general flow of the reconstruction process using deep learning algorithms in a hospital or clinical environment is illustrated in Fig. 7.14.

7.6 Conclusion

Image reconstruction is the heart of the CT scanner, where images of the patient are constructed from various projection datasets that contain attenuation values obtained at various angles as the X-ray tube and detectors rotate 360° around the patient. These algorithms include the FBP algorithm, IR algorithms, and more currently, deep learning algorithms based on AI techniques. It is important to realize that the evolution and development of these algorithms through the years are intended to benefit the patient through the use of the lowest dose possible without compromising image quality needed for diagnosis. The number of papers describing the use of

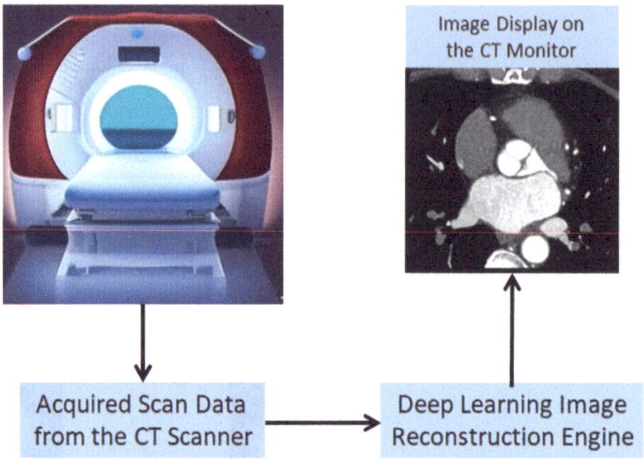

Fig. 7.14 The general flow of the reconstruction process using deep learning algorithms in a hospital or clinical environment. The CT scanner was produced using an AI image generator. (Image used with permission of GE HealthCare Technologies Inc.)

DL image reconstruction is increasing. In conclusion of this chapter, Arndt et al. [13] describing the use of DL CT image reconstruction state the following: "In our view machine learning will inevitably affect many aspects of radiology positively. DL image reconstruction demonstrates improved image quality in terms of denoising and artifact reduction and shows potential for dose optimization. However, the implementation is not straightforward and currently in early clinical stages while the explainability and reliability of machine learning are a focus of research. While early research findings are promising, DL image reconstruction should be further introduced into clinical practice and extensively investigated to provide evidence of clinical superiority" [13].

References

1. Najjar (2023). Redefining Radiology: A Review of Artificial Intelligence Integration in Medical Imaging. Diagnostics 2023, 13, 2760. Creative Commons Attribution (CC BY) license (https://creativecommons.org/licenses/by/4.0/).
2. Seeram E (2018). Computed tomography: a technical review. Radiol Technol;89(3):279–302.
3. Seeram E (2023). Computed Tomography: Physical Principles, Patient Care, Clinical Applications, and Quality Control. St Louis, MO. Elsevier.
4. Bushberg JT, Seibert JA, Leidholdt EM, Boone JM. Computed tomography. In: The Essential Physics of Medical Imaging. 4th Edition. Philadelphia, PA: Lippincott Williams & Wilkins; 2020:312–374.
5. Kaza RK, Platt JF, Goodsitt MM, et al. (2014). Emerging techniques for dose optimization in abdominal CT. Radiogr;34(1):4–17. https://doi.org/10.1148/rg.341135038
6. Herman GT (1980). Fundamentals of Computerized Tomography: Image Reconstruction from Projections. New York, New York; Academic Press.
7. Hsieh J, Nett B, Yu Z, Sauer K, Thibault J-B, Bouman CA (2013). Recent advances in CT image reconstruction. Curr Radiol Rep;1(1):39–51. https://doi.org/10.1007/s40134-012-0003-7

8. Hsieh J, Liu E, Nett B, Tang J, Thibault J-B, Sahney S (2019). A new era of image reconstruction: TrueFidelity. https://www.gehealthcare.ru//jssmedia/040dd213fa89463287155151fdb01922.pdf. Accessed February 14, 2024
9. Yaqub M, Jinchao F, Arshid K, Ahmed S, Zhang W, Nawaz MZ, Mahmood T (2022) "Deep Learning-Based Image Reconstruction for Different Medical Imaging Modalities", Computational and Mathematical Methods in Medicine; 8750648, https://doi.org/10.1155/2022/8750648
10. Koetzier LR, Mastrodicasa D, Szczykutowicz TP, van der Werf NR, Wang AS, Sandfort V, van der Molen AJ, Fleischmann D, Willemink MJ (2023). Deep Learning Image Reconstruction for CT: Technical Principles and Clinical Prospects. Radiology;306(3):e221257. https://doi.org/10.1148/radiol.221257. Epub 2023 Jan 31. PMID: 36719287; PMCID: PMC996877.
11. Zhang, M., Gu, S. & Shi, Y. (2022). The use of deep learning methods in low-dose computed tomography image reconstruction: a systematic review. Complex Intell. Syst; 8, 5545–5561, https://doi.org/10.1007/s40747-022-00724-7
12. Seibert JA (2014). Iterative reconstruction: how it works, how to apply it. Pediatr Radiol; 44(suppl 3):431–439. https://doi.org/10.1007/s00247-014-3102-1
13. Arndt C, Güttler F, Heinrich A etal. (2021). Deep Learning CT Image Reconstruction in Clinical Practice. Fortschr Röntgenstr; 193: 252–261
14. Ehman EC, Yu L, Manduca A, et al. (2014). Methods for clinical evaluation of noise reduction techniques in abdominopelvic CT. Radiogr;34(4):849–862. https://doi.org/10.1148/rg.344135128
15. Beister M, Kolditz D, Kalender WA (2012). Iterative reconstruction methods in X-ray CT. Phys Med;28(2):94–108. https://doi.org/10.1016/j.ejmp.2012.01.003
16. Scientific modelling. Science Learning Hub website. https://www.sciencelearn.org.nz/resources/575-scientific-modelling Accessed February 15, 2024.
17. Nagayama Y, Sakabe D, Goto M, Emoto T, Oda S, Nakaura T (2021). Deep Learning–based Reconstruction for Lower-Dose Pediatric CT: Technical Principles, Image Characteristics, and Clinical Implementations. RadioGraphics; 41:1936–1953
18. Qiu D, Seeram E (2016) Does iterative reconstruction improve image quality and reduce dose in computed tomography? Radiol Open J;1(2):42–54. https://doi.org/10.17140/ROJ-1-108
19. Sun, J., Li, H., Wang, B. et al. (2021). Application of a deep learning image reconstruction (DLIR) algorithm in head CT imaging for children to improve image quality and lesion detection. BMC Med Imaging; 21, 108. https://doi.org/10.1186/s12880-021-00637-w
20. Greffier J, Dabli D, Hamard A, Belaouni A, Akessoul P, Frandon J, Beregi JP (2022). Effect of a new deep learning image reconstruction algorithm for abdominal computed tomography imaging on image quality and dose reduction compared with two iterative reconstruction algorithms: a phantom study. Quant Imaging Med Surg;12(1):229–243. https://doi.org/10.21037/qims-21-215. PMID: 34993074; PMCID: PMC8666764.
21. Boedeker K (2019). AiCE deep learning reconstruction: bringing the power of ultra-high resolution CT to routine imaging. Canon Med Syst; 2:28–33.
22. What's that noise? How deep learning can elevate CT image quality, reduce dose exposure, and extend the life of older scanners. AlgoMedica https://cdn.b12.io/client_media/MylCrHXR/cd6b3270-78bd-11ec-aeee-0242ac110003-PixelShine_White_Paper_final_1.18.2021.pdf. Accessed 28 Feb 2022.

Chapter 8
Computer-Aided Detection/ Computer-Aided Diagnosis

8.1 Introduction

In a textbook entitled *Expert Radiology: Breast Imaging*, authors Birdwell and Castellino [1] state that "the terms computer-aided detection and computer-aided diagnosis are often used to describe the same process and, to confound the problem, they share the same acronym, CAD".

In an earlier paper on "Notable CAD papers published in Medical Physics" to celebrate the 50th Anniversary Papers on this topic, Giger et al. [2] defined CAD as "a diagnosis made by a radiologist who uses the output from a computer analysis of the image data in their decision-making process. The final medical decision is made by the radiologist, not the computer. Note that with CAD, the role of the computer analysis is not to replace the radiologist but rather to aid the radiologist in his/her image interpretation and/or decision making." Furthermore, radiologists used the output from the CAD system to provide a "second opinion" in image interpretation, as illustrated in Fig. 8.1.

8.2 The Birth of CAD: CADe and CADx

The notion of using the processing output of the computer to assist the radiologist (as compared to using the computer to interpret images) was the focus of research in the mid-1980s, at the Kurt Rossmann Laboratories for Radiologic Image Research in the Department of Radiology at the University of Chicago [3]. CAD was subsequently employed in clinical practice in mammography in detection breast cancer and in lesion detection in chest images [2]. For more details on the historical development of CAD, interested readers should refer to the paper by Doi [3].

E. Seeram, V. Kanade, *Artificial Intelligence in Medical Imaging Technology*,
https://doi.org/10.1007/978-3-031-64049-0_8

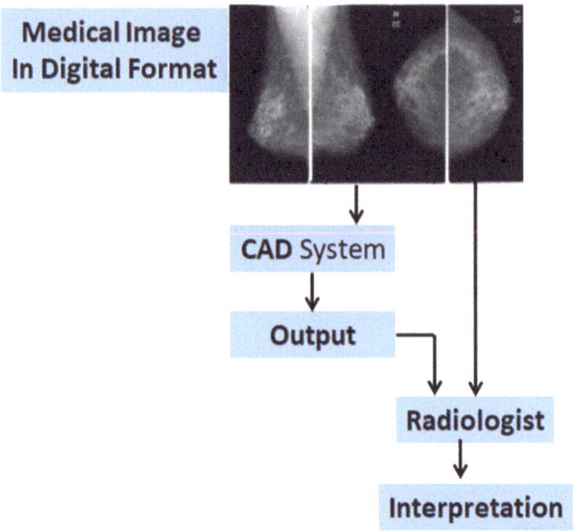

Fig. 8.1 Schematic diagram of a CAD system for medical image interpretation, where radiologists used the system to provide a "second opinion" in image interpretation. (From Giger ML, Chan H-P, Boone J (2008). Anniversary Paper: History and status of CAD and quantitative image analysis: The role of Medical Physics and AAPM; 35 (12); 5799–5820. Reprinted with permission under the Creative Commons Attribution 4.0 International (CC BY 4.0) license; https://creativecommons.org/licenses/by/4.0/)

Later, CAD evolved into two types, namely, computer-aided detection (CADe) and computer-aided diagnosis (CADx). While CADe "is intended to reduce the possibility of missing lesions by identifying abnormal regions of the image, CADx is intended to provide assistance to the human observer in evaluating various pathologies, based on the principles of classification, with the goal of automating the classification of abnormalities" [4].

The introduction of artificial intelligence in medical imaging has led to the evolution and development of more efficient CAD systems based on deep learning algorithms. This will be described in Sect. 8.4.

8.3 The Main Stages of a CAD System

An overview of the major components of an early CAD system is illustrated in Fig. 8.2. The system uses a number of steps that are trained to search for various patters and features that are suspicious and provides an output for review by, generally, a radiologist [5]. These include pre-processing, definition of a region of interest, feature extraction and selection, and classification.

A more detailed illustration of these steps carried out in a CAD system for breast cancer detection, and classification is shown in Fig. 8.3. These steps include pre-processing, mass segmentation, feature extraction, and classification [6]. It is not within the scope of this chapter to outline the details of each of these operations,

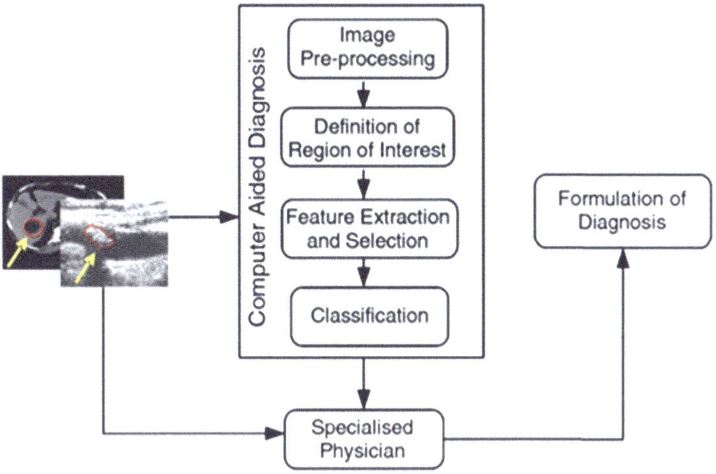

Fig. 8.2 The major stages characteristic of a computer-aided diagnosis (CAD) system in which the computer out acts as "second opinion" in image interpretation. (From Stoitsis J, Valavanis I, Mougiakakou SG, Golemati S, Nikita A, Konstantina S. Nikita KS (2006) Computer aided diagnosis based on medical image processing and artificial intelligence methods, Nuclear Instruments and Methods in Physics Research Section A: Accelerators, Spectrometers, Detectors and Associated Equipment; 569 (2); 591–595, https://doi.org/10.1016/j.nima.2006.08.134. Reproduced by permission)

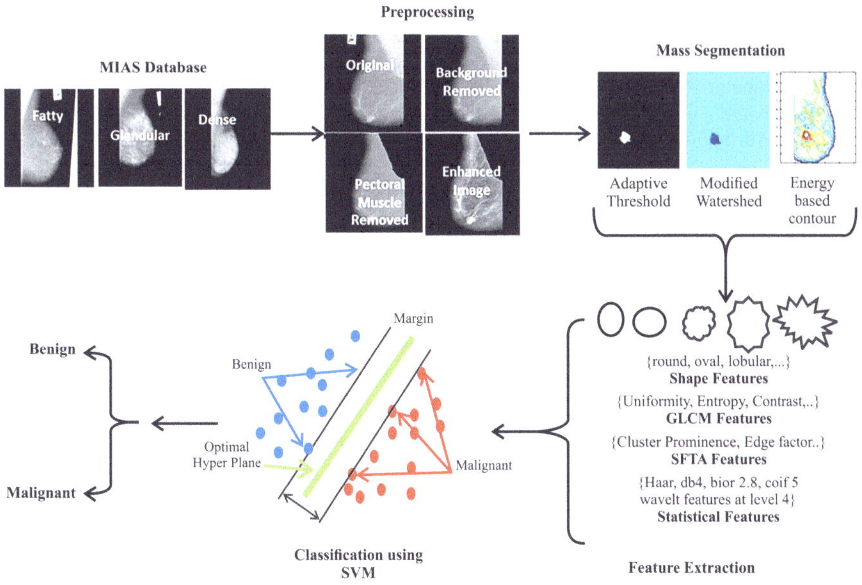

Fig. 8.3 A more detailed illustration of these steps carried out in a CAD system for breast cancer detection, and classification. (From Halalli, Bhagirathi, and Aziz Makandar. 2018. 'Computer Aided Diagnosis—Medical Image Analysis Techniques'. Breast Imaging. InTech. https://doi.org/10.5772/intechopen.69792. Reprinted with permission under the Creative Commons Attribution 4.0 International (CC BY 4.0) license https://creativecommons.org/licenses/by/4.0/)

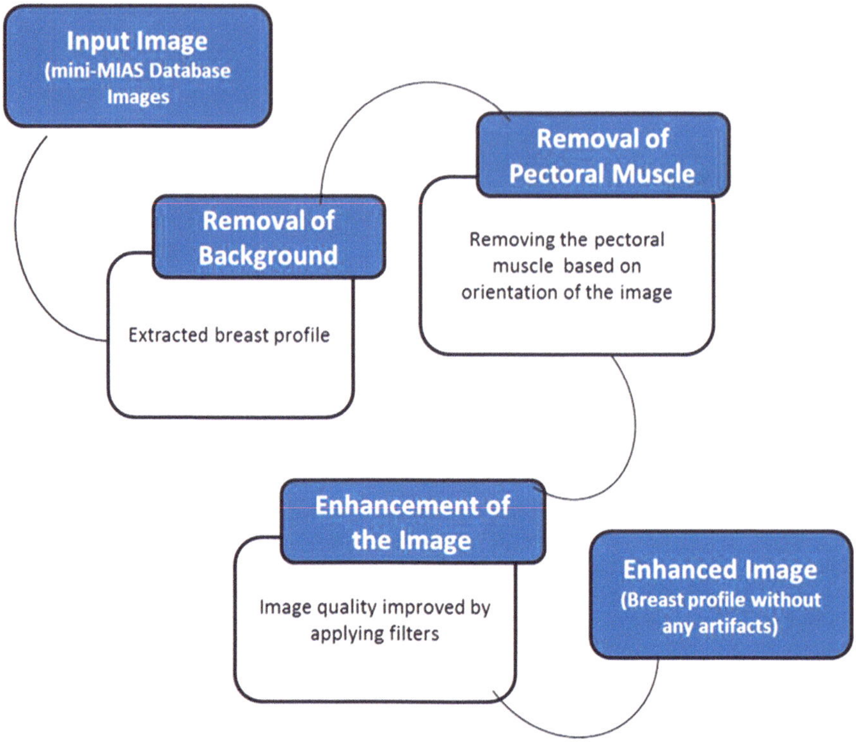

Fig. 8.4 The steps carried out in pre-processing. (From Halalli, Bhagirathi, and Aziz Makandar. 2018. 'Computer Aided Diagnosis—Medical Image Analysis Techniques'. Breast Imaging. InTech. https://doi.org/10.5772/intechopen.69792. Reprinted with permission under the Creative Commons Attribution 4.0 International (CC BY 4.0) license https://creativecommons.org/licenses/by/4.0/)

since they were described in detail in Chap. 5; however, the following points from Bhagirathi and Makandar [7] are noteworthy:

- Pre-processing, for example, includes several operations illustrated in Fig. 8.4, including removal of background, removal of pectoral mass, and image enhancement.
- Mass segmentation is used to separate out abnormal and normal portions of breast tissues. The overall goal of segmentation in the CAD model shown in Fig. 8.3 is mass segmented from the breast tissue as shown in Fig. 8.5 Mass in mammography is one of the subjects to identify the abnormality. Earlier mass detection operations [9–11] have been replaced by more recently used approaches such as region growing, watershed, threshold based, contour based, and clustering algorithms [7]. An illustration of the contour-based technique that works well on all kinds of tissues such as fatty, glandular, and dense and high-intensity and low-intensity images is shown in Fig. 8.6.

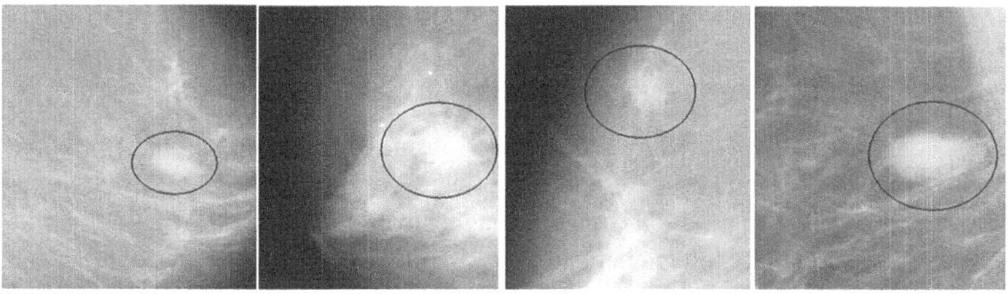

Fig. 8.5 The overall goal of segmentation in the CAD model is mass segmented from the breast tissue. (From Halalli, Bhagirathi, and Aziz Makandar. 2018. 'Computer Aided Diagnosis—Medical Image Analysis Techniques'. Breast Imaging. InTech. https://doi.org/10.5772/intechopen.69792. Reprinted with permission under the Creative Commons Attribution 4.0 International (CC BY 4.0) license https://creativecommons.org/licenses/by/4.0/)

(a)

(b)

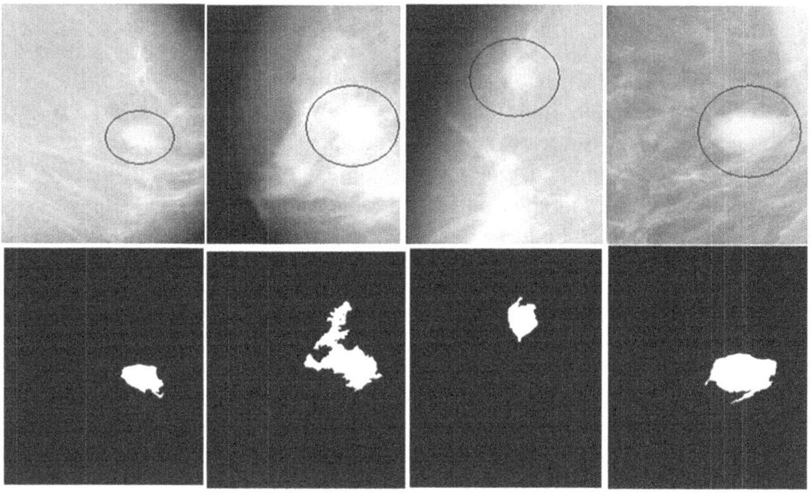

Fig. 8.6 Mass segmented on different tissues using contour-based segmentation. (**a**) Ground truth (**b**) and results of the proposed work. The term "ground truth" refers to the target for training or validating the model with a labeled dataset. (From Halalli, Bhagirathi, and Aziz Makandar. 2018. Computer Aided Diagnosis—Medical Image Analysis Techniques' Breast Imaging. InTech. https://doi.org/10.5772/intechopen.69792. Reprinted with permission under the Creative Commons Attribution 4.0 International (CC BY 4.0) license https://creativecommons.org/licenses/by/4.0/)

- Feature extraction refers to the extraction of mass shapes (Fig. 8.3) which changes from benign to malignant as round, oval, lobular, or irregular, using various operations such as wavelet features, gray level co-occurrence matrix (GLCM) features, and segmentation-based fractal texture analysis (SFTA) features [7]. These operations are not within the scope of this chapter; hence, the interested reader should refer to the paper by Bhagirathi and Makandar [7].

- Classification refers to the use of different algorithms such as the support vector machine (SVM) (Fig. 8.3), a supervised learning technique to classify suspicious regions of the tissues as benign or malignant [7].

In conclusion of this section, AlHinai [8] points out that "CAD only provides quantitative analysis and is consistently applied in routine or repetitive tasks. Signal analysis alone is not enough to make a diagnosis and needs to be integrated with other information that a physician can include such as the general physical appearance and mental state of the patient and family history. Accordingly, it is further emphasized that computer-aided diagnosis is used as a secondary source of diagnosis. The physicians are still the primary source of diagnosis. The CAD system will be used to provide a second opinion" [8].

8.4 Artificial Intelligence and CAD

Artificial intelligence (AI) has found applications in medical imaging modalities such as mammography, computed tomography, diagnostic radiography, nuclear medicine, and magnetic resonance imaging and in hybrid imaging such as single-photon emission computed tomography-computed tomography (SPECT/CT) and positron emission tomography-computed tomography (PET/CT) [12].

AI is now being applied to extend the usefulness and accuracy of conventional CAD systems used in breast imaging [4, 12–15]. Currently, the Food and Drug Administration in the United States has approved more than 20 AI applications for breast imaging [15–17]. The research on AI applications in breast imaging is gaining widespread attention, and the major focus of these projects is centered around cancer detection for mammography. Furthermore, other research focus on "decision support, risk assessment, breast density quantitation, workflow and triage, quality evaluation, response to neoadjuvant chemotherapy assessment, and image enhancement" [13].

8.4.1 Deep Learning-Based CAD Systems

The limitations of traditional CAD systems [4] have provided the motivation to develop new CAD platforms that will differ in many ways that of the traditional CAD platforms. These new CAD platforms will be based on deep learning (DL) algorithms. In DL, the use of artificial neural networks such as the convolutional neural network (CNN) is commonplace for image/object recognition and classification. Different types of neural networks were described in detail in Chap. 4. As an example of the use of DL-based CAD system on the variable of clinical utility, Gao et al. [4] indicate that while traditional CAD systems are limited and can detect ductal carcinoma in situ more than invasive ductal carcinoma, DL-based CAD will be "potentially unlimited" and may detect more invasive cancers.

An integrated DL-based CAD system is shown in Fig. 8.7. There are several steps to note as follows [14]:

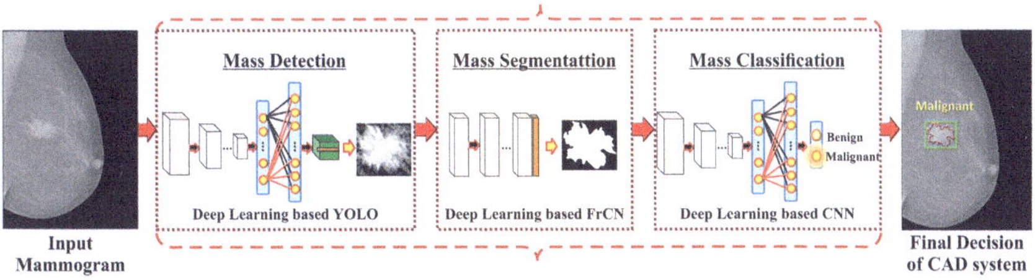

Fig. 8.7 The features of an integrated deep learning CAD system for breast lesion detection, segmentation, and classification. (From Mugahed A. Al-antari, Mohammed A. Al-masni, Mun-Taek Choi, Seung-Moo Han, Tae-Seong Kim (2018). A fully integrated computer-aided diagnosis system for digital X-ray mammograms via deep learning detection, segmentation, and classification, International Journal of Medical Informatics, Volume 117, 2018, Pages 44–54, https://doi.org/10.1016/j.ijmedinf.2018.06.003. Reproduced by permission)

- Training of the DL model is required to initialize the system.
- Three components are essential in order to provide a useful output to assist with decision-making and accurate diagnosis, namely, mass detection, mass segmentation, and mass classification. These architectures were described in detail in Chap. 4.
- For object detection (e.g., breast lesions), an algorithm called You Only Look Once (YOLO) is used.
- For mass segmentation of breast lesions, full resolution convolutional network (FrCN) is used not only to retain fine structural details but also to potentially improve the accuracy of detection, classification, and segmentation of medical conditions, leading to more reliable and precise medical assessments, and predictive analyses, as explained in Chap. 4.
- The final step involves the use of three DL architectures, namely, regular feedforward CNN: ResNet-50 and InceptionResNet-V2, for classification and recognition of lesions that are either benign or malignant [14]. These architectures were described in Chap. 4. In review, a feedforward neural network is one of the simplest and refers to the data moving in one direction only, that is, forward, as illustrated in Fig. 8.8, from the input to the convolutional and pooling layers and fully connected onto the output. No iteration is involved. ResNet-50 is a 50-layer convolutional neural network. Lastly, the Inception-ResNet-v2 is a DL convolutional neural network (164 layers deep) that is trained on more than a million images from the ImageNet database [18].

The use of DL-based CAD systems will not depend on the radiologist, and these systems will learn and improve with more and more datasets [15]. In this regard, "it will be necessary in the future to define the roles of radiologists…..These roles may change over time but in the near future may include tasks that are best served by AI alone, radiologists alone, or AI and radiologists together" [15], as illustrated in Fig. 8.9.

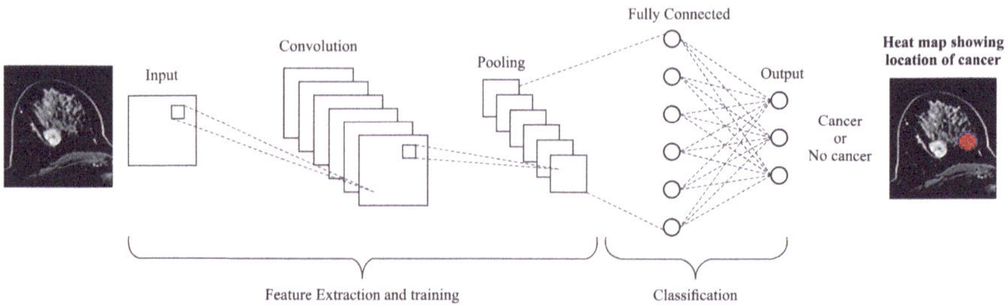

Fig. 8.8 Typical architecture of a convolutional neural network (CNN). The input layer represents the breast cancer image that serves as input to the CNN. The multiple convolutional layers are stacked on top of the input layer. Pooling layers are used to downsample the spatial dimensions of the feature maps. Fully connected layers are then employed to connect all the neurons from the previous layers to the subsequent layers. The final layer is the output layer, which provides the classification (Figure and legend From Adam R, Dell'Aquila K, Hodges, L. et al. (2023). Deep learning applications to breast cancer detection by magnetic resonance imaging: a literature review. Breast Cancer Res 25, 87. https://doi.org/10.1186/s13058-023-01687-4. Reprinted with permission under the Creative Commons Attribution 4.0 International (CC BY 4.0) license https://creativecommons.org/licenses/by/4.0/)

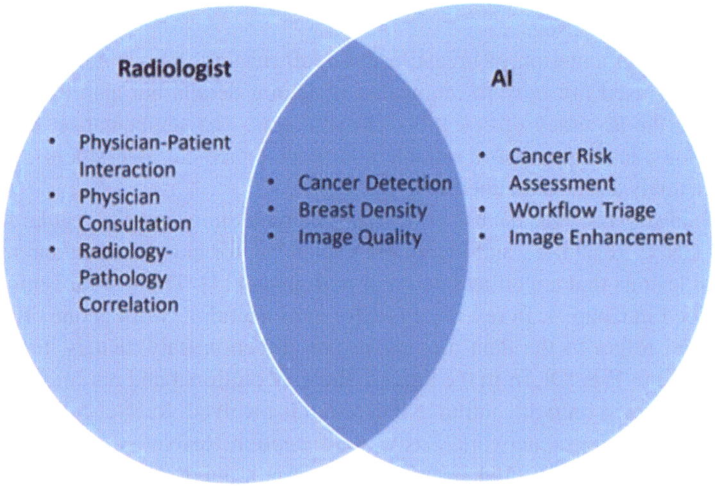

Fig. 8.9 Venn diagram demonstrating possible future situations where radiologists or AI can make contributions separately and together toward breast imaging care. (Figure and legend From Taylor, Clayton R., Natasha Monga, Candise Johnson, Jeffrey R. Hawley, and Mitva Patel. 2023. "Artificial Intelligence Applications in Breast Imaging: Current Status and Future Directions" Diagnostics 13, no. 12: 2041. https://doi.org/10.3390/diagnostics13122041. Reprinted with permission under the Creative Commons Attribution 4.0 International (CC BY 4.0) license https://creativecommons.org/licenses/by/4.0/)

References

1. Birdwell RL, Castellino RA (2011) Chapter 5: Computer-Aided Detection, Bassett LW, Mahoney, MC, Apple SK, D'Orsi CJ (Editors); In Expert Radiology, Breast Imaging. Philadelphia, PA, W.B. Saunders, pages 99–109; ISBN 9781416051992, https://doi.org/10.1016/B978-1-4160-5199-2.00005-9. (https://www.sciencedirect.com/science/article/pii/B9781416051992000059
2. Giger ML, Chan H-P, Boone J (2008). Anniversary Paper: History and status of CAD and quantitative image analysis: The role of Medical Physics and AAPM; 35 (12); 5799–5820
3. Doi K (2007) Computer-aided diagnosis in medical imaging: historical review, current status and future potential. Comput Med Imaging Graph; 31(4–5):198–211. https://doi.org/10.1016/j.compmedimag.2007.02.002. Epub 2007 Mar 8. PMID: 17349778; PMCID: PMC1955762.
4. Gao Y, Geras KJ, Lewin AA, Moy L (2019). New Frontiers: An Update on Computer-Aided Diagnosis for Breast Imaging in the Age of Artificial Intelligence. AJR Am J Roentgenol; 212(2):300–307. https://doi.org/10.2214/AJR.18.20392. Erratum in: AJR Am J Roentgenol; 212(3):712. PMID: 30667309; PMCID: PMC6927034.
5. Stoitsis J, Valavanis I, Mougiakakou SG, Golemati S, Nikita A, Konstantina S. Nikita KS (2006) Computer aided diagnosis based on medical image processing and artificial intelligence methods, Nuclear Instruments and Methods in Physics Research Section A: Accelerators, Spectrometers, Detectors and Associated Equipment; 569 (2); 591–595, https://doi.org/10.1016/j.nima.2006.08.134.
6. Guo, Z., Xie, J., Wan, Y., Zhang, M., Qiao, L., Yu, J., Chen, S., Li, B. & Yao, Y. (2022). A review of the current state of the computer-aided diagnosis (CAD) systems for breast cancer diagnosis. Open Life Sciences, 17(1), 1600–1611. https://doi.org/10.1515/biol-2022-0517
7. Bhagirathi H, Makandar A (2018). 'Computer Aided Diagnosis – Medical Image Analysis Techniques'. Breast Imaging. InTech. https://doi.org/10.5772/intechopen.69792
8. AlHinai N (2020) Introduction to biomedical signal processing and artificial intelligence. in Biomedical Signal Processing and Artificial Intelligence in Healthcare. St Louis, MO/Cambridge, MA; Elsevier-Academic Press
9. Cascio DO, Fauci F, Magro R, Raso G, Bellotti R, De Carlo F, Tangaro S, De Nunzio G, Quarta M, Forni G, Lauria A (2006). Mammogram segmentation by contour searching and mass lesions classification with neural network. IEEE Transactions on Nuclear Science; 53(5):2827–2833
10. Hassan SA, Sayed MS, Farag F (2014). Segmentation of breast cancer lesion in digitized mammogram images. Biomedical Engineering Conference (CIBEC); Cairo International, Giza, Egypt. IEEE.
11. Anitha J, Dinesh Peter J, Immanuel Alex Pandian S (2017). A dual stage adaptive thresholding (DuSAT) for automatic mass detection in mammograms. Computer Methods and Programs in Biomedicine; 138:93–104
12. Najjar R (2023). Redefining Radiology: A Review of Artificial Intelligence Integration in Medical Imaging. Diagnostics 13, no. 17: 2760. https://doi.org/10.3390/diagnostics13172760
13. Koenigkam-Santos M, Ferreira-Júnior JR, Wada DT, Tenório APM, Nogueira-Barbosa MH, Azevedo-Marques PMA (2019). Artificial intelligence, machine learning, computer-aided diagnosis, and radiomics: advances in imaging towards to precision medicine. Radiol Bras; 52(6):387–396.
14. Mugahed A. Al-antari, Mohammed A. Al-masni, Mun-Taek Choi, Seung-Moo Han, Tae-Seong Kim (2018). A fully integrated computer-aided diagnosis system for digital X-ray mammograms via deep learning detection, segmentation, and classification, International Journal of Medical Informatics, Volume 117, 2018, Pages 44–54, https://doi.org/10.1016/j.ijmedinf.2018.06.003.
15. Taylor CR, Monga N, Johnson C, Hawley JR, Patel M (2023). Artificial Intelligence Applications in Breast Imaging: Current Status and Future Directions. Diagnostics 13, no. 12: 2041. https://doi.org/10.3390/diagnostics13122041

16. Adam R, Dell'Aquila K, Hodges, L. et al. (2023). Deep learning applications to breast cancer detection by magnetic resonance imaging: a literature review. Breast Cancer Res 25, 87. https://doi.org/10.1186/s13058-023-01687-4
17. Kadhim YA, Khan MU, Mishra A (2022). Deep Learning-Based Computer-Aided Diagnosis (CAD): Applications for Medical Image Datasets, Sensors; 22, no. 22: 8999. https://doi.org/10.3390/s22228999
18. Inceptionresnetv2: Pretrained Inception-ResNet-v2 convolutional neural network. https://www.mathworks.com/help/deeplearning/ref/inceptionresnetv2.html Accessed March 1st, 2024

Chapter 9
Ethical and Regulatory Considerations

9.1 Introduction

Ethics, rooted in the philosophical exploration of morality and decision-making, serves as the guiding compass for human actions and behaviors. Renowned philosopher Aristotle emphasized ethics as the pursuit of virtuous living, stating that "ethics is not just about knowing, but about doing the good" [1]. Similarly, Immanuel Kant, a pivotal figure in ethical theory, proposed the categorical imperative, highlighting the universalizability of ethical principles: "Act only according to that maxim whereby you can, at the same time, will that it should become a universal law" [2]. In essence, ethics delves into the principles and values that govern human conduct, navigating the complex terrain of right and wrong, guiding individual and societal choices.

In the field of AI in medical imaging, ethical and regulatory considerations stand as pivotal pillars for the responsible integration of AI technologies. Ethical dimensions highlight the necessity for radiology AI to prioritize well-being, minimize harm, and equitably distribute benefits and risks among stakeholders. Central ethical issues involve preserving human autonomy, preventing harm, ensuring fairness and transparency, and enhancing the interpretability of AI systems. The advent of AI in medical imaging introduces ethical dilemmas encompassing patient privacy, potential unintended harm, equity for marginalized groups, and transparency in clinical performance.

Concurrently, regulatory aspects play a crucial role, notably in aligning with data protection regulations such as Europe's General Data Protection Regulation (GDPR) [3]. Bodies such as the Food and Drug Administration (FDA) have instituted specific processes and considerations for AI/ML technologies within medical imaging devices [4–10]. The WHO emphasizes the need for ensuring AI systems' safety and effectiveness, fostering collaboration among regulatory entities, healthcare

E. Seeram, V. Kanade, *Artificial Intelligence in Medical Imaging Technology*,
https://doi.org/10.1007/978-3-031-64049-0_9

professionals, and industry representatives while navigating intricate data protection regulations [11].

Legal considerations highlight that AI algorithms used in medicine fall under FDA-regulated medical devices, fueling ongoing discussions to define comprehensive policies and frameworks to address the legal and ethical facets of AI in healthcare, including medical imaging.

Ultimately, ethical and regulatory considerations are fundamental to ensuring the responsible and secure utilization of AI technologies in medical imaging.

9.2 Regulatory Frameworks and Guidelines for AI in Healthcare (e.g., FDA Regulations)

Artificial intelligence (AI) is becoming increasingly fundamental in the healthcare sector, promising to transform patient care and healthcare delivery by improving the accuracy and efficiency of diagnosis and treatment. Yet, this revolutionary technology also elevates questions around ethical considerations, data privacy, and patient safety.

Comprehensive regulatory frameworks and guidelines are instrumental to ensure these pioneering technologies are safe, effective, and equitable. Recent developments in regulatory frameworks, particularly in the United States and the United Kingdom, as well as global guidelines, illuminate the paths taken by authorities to address these matters.

United States: FDA Regulation

In the United States, the Food and Drug Administration (FDA) plays a pivotal role in regulating medical devices, including those embedded with AI and machine learning capabilities. The FDA evaluates AI-based medical devices through various pathways, including the pre-market approval (PMA) process or the 510(k) clearance pathway [6, 9].

- **510(k) clearance**: The 510(k) clearance pathway is a regulatory process implemented by the US FDA for medical devices. This pathway is designed for devices that are considered substantially equivalent to those that have already been cleared by the FDA [9].

When a manufacturer develops a new medical device that shares similarities with a device already on the market, they can apply for 510(k) clearance. This process involves demonstrating that the new device is similar enough in design, technology, and intended use to an existing FDA-cleared device. By proving this substantial equivalence, the new device can potentially be cleared for market entry without the need for extensive clinical trials.

- **Pre-market approval (PMA)**: For higher-risk devices or those without a substantial equivalent, PMA requires comprehensive clinical data demonstrating safety and effectiveness [6, 10].

Recognizing the dynamic nature of AI/ML technologies, the FDA released an action plan in April 2021 for AI/ML-based software as a medical device (SaMD) [5]. This initiative is outlined in the discussion paper titled "Proposed Regulatory Framework for Modifications to Artificial Intelligence/Machine Learning-Based Software as a Medical Device" [4]. The plan outlines a regulatory approach that accounts for the iterative nature of AI by anticipating modifications and adjusting the validation process accordingly.

As of October 2023, the FDA has greenlit 692 AI/ML-enabled medical devices, a staggering 80% of which received approval after 2019 [12, 13]. These approvals were granted through pathways such as 510(k) clearance, de novo requests, or pre-market approvals. Notably, as of the latest update, devices relying solely on a generative AI architecture haven't received FDA approval [12]. A comprehensive list of AI-/ML-enabled medical devices available in the United States is available at FDA's official catalog titled "Artificial Intelligence and Machine Learning (AI/ML)-Enabled Medical Devices" [12].

One of the core components of the FDA's regulations for AI/ML in healthcare focuses on the total product life cycle (TPLC) regulatory approach [14]. This method ensures compliance with safety and effectiveness standards from the initial development stages through post-market performance. The approach effectively positions the FDA to supervise the adaptive nature of AI which can constantly learn and evolve from new data.

Under these guidelines, developers are required to submit a "predetermined change control plan" with their initial product submission, detailing how they will manage changes to their algorithms, including those driven by ML. Additionally, a heavy emphasis is placed on real-world performance monitoring [15]. The FDA's guidance suggests enhancing transparency and trust by ensuring robust surveillance systems that could track the performance of AI applications once deployed in clinical settings.

Furthermore, the Health Insurance Portability and Accountability Act (HIPAA) in the United States, while not AI-specific, sets a precedent for protecting patient information. Also, it could provide a model for additional AI-focused regulatory measures in healthcare data security [16].

United Kingdom: MHRA's Standpoint

Across the Atlantic, the UK's Medicines and Healthcare products Regulatory Agency (MHRA) has engaged in a parallel endeavor. Recently, the MHRA released a guidance document stating that AI applications should be managed within the broader scope of medical device regulation under the UK law [17]. This means that the typical AI applications need to comply with the same requirements as traditional devices in terms of safety, quality, and performance. The UK's stance is firmly embedded in ensuring that AI/ML applications are validated with the same rigor that is applied to medical devices. MHRA also underscores the importance of clinical evaluation and post-market surveillance to track the real-world performance and potential impacts of the technology on patients.

Global Perspectives and Ethical Considerations

Globally, institutions such as the World Health Organization (WHO) are also contributing to the formulation of guidelines. In June 2021, WHO published a report, "Ethics and governance of artificial intelligence for health," which is the outcome of 2 years of consultations held by a panel of international experts [18]. This report is comprehensive, covering aspects such as protecting human autonomy, ensuring equity and inclusivity, promoting AI that is responsive and sustainable, and enhancing AI that is transparent and explainable.

Furthermore, the Global Partnership on AI (GPAI), an initiative launched in June 2020, established by 15 countries, including the United States and United Kingdom, aims to bridge the gap between theory and practice by supporting the responsible development and use of AI [19, 20]. This initiative complements the G7 nations' efforts in coordinating economic policies, among them, the regulation of AI in healthcare.

On the European front, the European Union (EU) has proposed the Artificial Intelligence Act (AI Act), which is currently in the legislative process [21]. Although it is yet to be finalized, the EU AI Act stands to be one of the most influential regulatory frameworks globally, given the gravitas of the EU in international trade and the precedent it sets for AI regulation which can extend beyond its borders.

The proposed EU AI Act is extensive and focuses on a risk-based approach to AI regulation, categorizing AI systems into unacceptable risk, high risk, limited risk, and minimal risk. For healthcare, which is principally categorized under the high-risk bracket, the AI Act would impose strict obligations on transparency, data governance, and human oversight. Provisions in the AI Act emphasize that users of high-risk AI systems should be provided with information on the system's capabilities and limitations, ensuring transparency in how AI systems work within healthcare contexts.

Furthermore, the MHRA, along with the US FDA and Health Canada, has come together to outline ten key principles that can guide the development of Good Machine Learning Practice (GMLP) [22]. They have also collaboratively established five principles for using predetermined change control plans (PCCPs). This ensures that our future guidance on PCCPs aligns internationally with these regulatory bodies [23].

Industry-Specific Initiatives

From the perspective of industry-specific guidelines, notable frameworks such as the FDA's Digital Health Software Precertification (Pre-Cert) Program propose a new way forward [24]. The Pre-Cert pilot program, which includes major players such as Apple and Fitbit, aims to develop a more streamlined and efficient approach to regulatory assessments [25]. This mechanism indicates a shift in the regulatory mindset from solely product-based evaluation to include assessment of the entity behind the technology, which is particularly relevant in the rapidly evolving realm of AI.

Standard-Setting Organizations

Apart from regulatory institutions, standard-setting organizations such as the International Organization for Standardization (ISO) and the Institute of Electrical and Electronics Engineers (IEEE) are developing standards that address various aspects of AI, such as ethical design (IEEE P7000 series) and AI application (ISO/IEC JTC 1/SC 42), which are essential for AI systems in healthcare [26, 27]. These standards often go hand in hand with the regulatory frameworks set by bodies such as the FDA, aiming to create a cohesive environment for AI in healthcare.

The IEEE P7000 series, for instance, tackles the challenge of value alignment, ensuring that AI systems perform tasks in ways that are aligned with human values and ethical considerations [28]. The series includes various standards focusing on distinct aspects of the ethical challenges posed by AI systems, such as transparency (IEEE P7001), accountability (IEEE P7002), and personal data protection (IEEE P7003).

Similarly, ISO/IEC JTC 1/SC 42 stands for ISO, International Organization for Standardization; IEC, International Electrotechnical Commission JTC 1; and Joint Technical Committee 1 SC 42, Subcommittee 42. The subcommittee is specifically focused on the standardization of AI and related technologies. This encompasses AI and automation processes, contributing to the establishment of a universal language and quality benchmarks for product development and evaluation in healthcare AI [27].

These initiatives reflect a growing recognition that AI's potential in healthcare will be best realized within a framework that promotes ongoing learning and improvement while ensuring that such systems are trustworthy and aligned with public health goals. The combined efforts of regulatory agencies and standard-setting organizations are pivotal to shaping the utilization of AI in healthcare and ensuring that advancements in this field are beneficial, ethical, and sustainable.

9.3 Privacy and Security Considerations for Medical Imaging Data

In the realm of medical imaging, ensuring the privacy and security of patient data is paramount. As technology advances and healthcare becomes increasingly digitized, the protection of sensitive medical information contained within imaging datasets is crucial. This involves navigating intricate privacy regulations and implementing stringent security measures to safeguard patient confidentiality while maintaining the integrity of the data [29].

Let's now explore the essential privacy and security aspects linked to medical imaging data (Fig. 9.1).

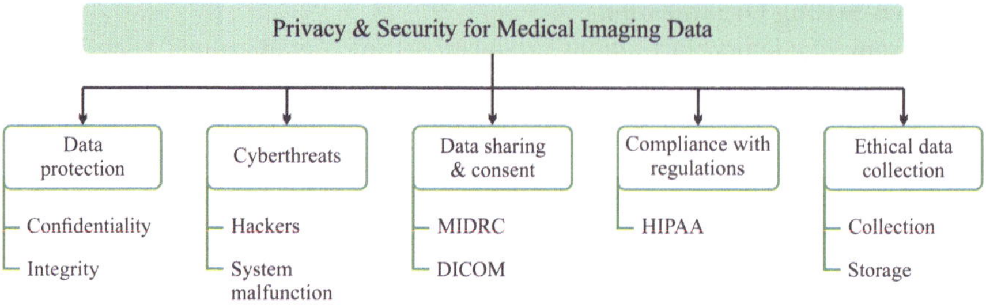

Fig. 9.1 Privacy and security aspects of medical imaging data. See text for further explanation

1. **Data Protection**

Medical imaging data consists of highly detailed images such as X-rays and MRIs, which reveal intimate details about a person's internal health state. Alongside these images, oftentimes, there's also an abundance of associated metadata, which can include everything from the patient's identifying information to their medical history.

The confidentiality, integrity, and availability of this medical imaging data are governed by strict regulatory frameworks, such as the Health Insurance Portability and Accountability Act (HIPAA) in the United States or the General Data Protection Regulation (GDPR) in Europe [3, 16]. These regulations mandate both the protection of privacy and the safeguarding of data against unauthorized access or breaches.

To protect patient data, the following measures can help:

- **Restrict access**: Authorized personnel should be the only ones to access medical imaging data. Access control mechanisms, such as role-based access control (RBAC), should be used to enforce data privacy within the information system application [30].
- **Strong passwords and multifactor authentication**: These measures add an extra layer of security to protect accounts and sensitive health data [31].
- **Regular updates and backups**: Keeping the operating system, software, and firmware of medical devices up-to-date can help fix known vulnerabilities and protect against the latest cyber threats [32]. Regular backups of patient data should also be in place to ensure its availability in case of a disaster.
- **Encryption**: Encrypting medical data can help protect it from unauthorized access and maintain its confidentiality [33].
- **Anonymization and de-identification**: Anonymizing and de-identifying data in medical imaging involves removing or coding specific information that could identify a patient. This process is crucial for protecting patient privacy. However, it's becoming more challenging due to the growing number of exceptions in medical imaging data. These exceptions might include unique features or markers within images that could inadvertently reveal a patient's identity even after attempts to anonymize the data.

- **Staff training and awareness**: Healthcare providers should educate their staff about the importance of data security and privacy and provide regular training to ensure staff members are aware of potential risks and threats to patient data.

By implementing these measures, healthcare providers can help minimize the risks to patient data and ensure the secure storage and sharing of medical images.

2. **Cyber Threats**

Medical imaging data, when not properly secured, can be vulnerable to a variety of cyber threats:

- **Cyberattacks**: Hackers can target healthcare databases to steal or manipulate sensitive patient data [49, 50].
- **Human error**: Mistakes by healthcare staff can lead to accidental exposure or loss of data.
- **Insider threats**: Employees with access can occasionally abuse their data access for unauthorized purposes.
- **Technical failures**: System malfunctions or software glitches can lead to data corruption or loss.

The implications of such incidents are grave, from potential misdiagnoses to the loss of public trust and expensive legal battles. To address these challenges, healthcare providers and device manufacturers should follow cybersecurity best practices, such as the following:

- **Timely updates**: Keeping the medical devices up-to-date can help fix known vulnerabilities and protect against the latest cyber threats.
- **Collaboration with device manufacturers**: Healthcare providers should work closely with device manufacturers to ensure that medical devices are designed and manufactured with robust cybersecurity measures in place.
- **Monitoring and maintenance**: Cybersecurity must be monitored and maintained throughout the device's life cycle, as new vulnerabilities may be discovered and require patches or fixes for new cybersecurity risks [34].

With these best practices, healthcare providers can cut down on risks and threats to patient's sensitive health data and medical devices.

3. **Data Sharing and Consent**

Medical imaging data is increasingly used for research and development, necessitating the establishment of standards for data sharing and patient consent. Patients should be informed about the use of their data and have the right to opt-out or withdraw their consent. Firstly, the need for standardized data sharing and patient consent mechanisms is crucial. Initiatives such as the Medical Imaging and Data Resource Center (MIDRC) aim to create reliable repositories for medical images and associated clinical data [35].

The Digital Imaging and Communications in Medicine (DICOM) standard has significantly improved data exchange among researchers, creating value for medical research [36]. It offers the language that medical devices and systems use to share and process medical images and related information. Yet, the sharing of medical imaging data must strictly adhere to consent principles, ensuring patients voluntarily agree to share their clinical images and understand the associated benefits and risks.

Collaborative platforms such as GIFT-Cloud and organizations such as the Advanced Research Projects Agency for Health (ARPA-H) Biomedical Data Fabric (BDF) Toolbox underscore the growing collaboration between clinical institutions and research communities [37, 38]. This collaboration emphasizes both the sharing and responsible utilization of medical imaging data for research and development while upholding stringent privacy and security standards.

4. **Compliance with Regulations**

Compliance with regulations, notably frameworks such as the HIPAA in the United States, is a cornerstone for the implementation of AI in medical imaging. It mandates strict controls on the handling and disclosure of protected health information (PHI) [39]. These regulations are specifically designed to safeguard patient information, ensuring its confidentiality and security.

Noncompliance with these regulations could result in severe legal consequences for healthcare providers and organizations. Penalties for breaches of regulations such as HIPAA can include substantial fines, legal action, and damage to the institution's reputation.

Compliance not only mitigates legal risks but also fosters trust between patients and healthcare providers, assuring individuals that their personal health information is handled with the highest standards of privacy and accountability. Thus, the integration of AI in medical imaging necessitates strict adherence to regulatory frameworks such as HIPAA to maintain patient privacy and avoid potential legal repercussions.

5. **Ethical Data Collection**

The collection, storage, and sharing of medical imaging data should be carried out in an ethically guided manner. This includes avoiding any bias in the selection of training data, ensuring that the analysis is not discriminatory against certain groups or individuals, and adhering to established clinical and research practices for informed consent.

Best Practices for the Future

Going forward, for healthcare providers, adopting a security-by-design approach is crucial. This means incorporating security considerations at every step of the medical imaging process. It may include developing new secure technologies (such as blockchain, cloud services, etc.) to handle day-to-day patient data [34, 40]. Regular security assessments can provide early detection of potential vulnerabilities, allowing for proactive management.

At the policy level, the ongoing development and refinement of global and national standards for privacy and security are directly impacting how medical imaging data is processed and protected. These standards should be adaptable to the rapid evolution of both the healthcare and cyber landscapes.

Furthermore, embracing a culture of security among all members of a healthcare organization can significantly strengthen the protection of medical imaging data. Security should not be seen as just an IT issue but as a fundamental component of healthcare ethics and patient care.

In this drive to secure medical imaging data, it must be remembered that the ultimate goal is not only to protect bits and bytes but also to maintain the trust and confidence of patients and the public. The privacy and security of medical imaging data are not standalone technical issues but impinge directly upon physician-patient relationships and the integrity of medical diagnostics.

9.4 Bias, Fairness, and Transparency in AI Algorithms

The integration of artificial intelligence (AI) in medical imaging has marked a transformative era in healthcare diagnostics. AI algorithms increasingly assist in the interpretation of medical images such as X-rays, MRIs, and CT scans, leading to early disease detection, personalized treatment plans, management of multiple medical conditions, and overall improved patient care.

However, as the technology advances, so do the concerns surrounding the potential biases within these algorithms, and the subsequent implications on fairness, transparency, and trust. Understanding these concerns requires an analysis distinguished by nuanced examples and considerations of their impact.

Bias in AI Algorithms

Bias in AI algorithms occurs when an AI systematically and unfairly prefers certain individuals or groups over others. In medical imaging, this can manifest in numerous ways:

- **Data collection bias**: If the algorithm is trained exclusively on images from a particular demographic (e.g., a specific ethnicity or gender), it may not perform well on broader populations. As a unique example, a study that utilizes chest X-rays mostly from middle-aged white males to train a pneumonia detection tool may miss signs of the disease in women or people from other ethnic backgrounds.
- **Annotation bias**: Bias can also emerge during the labeling process of medical images. If radiologists annotated images for bone fracture detection but had varying interpretations of what constitutes a fracture (due to differences in training or subjective judgment), the AI model could inherit inconsistencies, leading to inaccuracies in automated fracture detection.
- **Algorithm design bias**: Consider an AI system developed to recognize signs of a rare disease but trained mainly on early-stage manifestations of the ailment. The model might fail to recognize more advanced stages, incorrectly assuring patients they are disease-free.

Fairness in AI Algorithms
Fairness in AI algorithms is about ensuring that an AI system provides equal and unbiased outcomes and decision-making for diverse groups. Here are two key aspects to understand:

- **Equal performance across demographics**: A diabetic retinopathy detection AI should perform with similar accuracy, regardless of the patient's location, whether in rural Africa or urban America.
  Significance: This isn't just about making the AI system efficient. It's actually about a broader ethical responsibility, ensuring that the AI performs similarly across different locations and aligns with the ethical commitment to granting everyone equal access to healthcare.
  By maintaining consistent accuracy regardless of location, the AI helps bridge potential healthcare gaps. It ensures that people, irrespective of where they live, have access to similar quality diagnostic capabilities, thus promoting fairness in healthcare services.
- **Algorithmic treatment suggestions**: An AI system recommending cancer treatments should not favor treatments based on a patient's socioeconomic status (revealed in patient's profile information or historical data), unless such information is directly relevant to treatment efficacy.

Significance: This is crucial because it prevents any unfair advantages or disadvantages in healthcare based on a person's wealth or social status. The significance lies in ensuring that the AI system makes treatment recommendations solely on medical effectiveness, rather than on someone's financial situation.

By avoiding favoritism related to the socioeconomic status, the AI contributes to fair and unbiased decision-making in medical treatments. It aims to provide equal opportunities for effective healthcare to everyone, regardless of their financial or social background, emphasizing the importance of fairness in healthcare practices.

Transparency in AI Algorithms
Transparency in AI involves the clarity and openness with which the working mechanisms and decision processes of an AI system are made known to users and stakeholders. Here are two key components:

- **Explainability of decisions**: In an AI model predicting the likelihood of malignant tumors, it should provide insights into why it flagged certain cases as high risk. If the AI system identifies a patient as having a high risk of cancer based on unusual features, medical professionals need to understand the rationale behind these predictions for informed decision-making.
  Significance: This ensures that those using or relying on AI-generated insights can comprehend the logic behind the system's decisions, fostering trust, and aiding healthcare professionals in their assessments.
- **Disclosure of limitations**: Companies creating AI diagnostic tools must be transparent about the limitations of their products. For instance, a neural network that excels in detecting blood clots in CT scans may struggle with scans from machines it wasn't trained on. Users need to be aware of these limitations to avoid potential misdiagnoses.

Significance: This transparency helps users, such as healthcare providers, understand the boundaries of the AI system, enabling them to use the technology appropriately and make well-informed decisions based on its capabilities and limitations.

Key Examples
Now let's illustrate these principles with specific examples:

1. **Bias**—An AI model for breast cancer screening may demonstrate high accuracy in urban hospitals that contribute data rich in diversity. However, if deployed in rural areas where the demographics and technology (such as mammogram equipment) differ, the AI could underperform due to its data collection bias. For instance, it may not identify as malignant those tumor characteristics more common in the rural population, which were underrepresented in the training set.
2. **Fairness**—In pathology, an AI designed to recognize various skin diseases should equally identify conditions across different skin tones. An example of fairness would be adjusting the data and model to ensure accurate diagnosis of melanoma regardless of whether it appears on light or dark skin, rectifying a historical bias where darker-skinned individuals have been underdiagnosed.
3. **Transparency**—An AI tool used for detecting strokes in neuroimaging should offer a user-friendly explanation interface that provides reasons for its findings, such as highlighting the infarcted areas it identifies on brain images. This transparency allows medical professionals to see the basis for AI decisions and verify their correctness [29].

To wrap up, it is critical for AI models in medical imaging to adhere to these guiding principles, as deviations can lead to misdiagnoses and the propagation of healthcare inequities. Moreover, it emphasizes the significance of diverse and inclusive datasets to mitigate bias and the importance of rigorous testing across populations to ensure fairness. Additionally, maintaining open channels of communication on how AI works is crucial to build trust and transparency. With ethical AI development paired with continuous evaluation and recalibration, AI can enhance medical imaging to benefit all patient groups fairly and openly.

9.5 Responsible Use of AI in Healthcare

Upon the integration of AI into healthcare, a critical examination of its impact on accountability and responsibility becomes essential. While AI is poised not only to match but also to outperform healthcare professionals, it is not infallible. When errors arise within AI systems, the question of accountability surfaces.

The effectiveness of AI heavily relies on the caliber of the medical dataset used for its training. For example, an AI system trained on top-notch data, such as detailed brain scans in a well-controlled, high-quality environment, might struggle when dealing with lower-quality data, such as blurry scans, or when functioning in less-than-ideal clinical environments with outdated equipment.

Moreover, the context in which AI algorithms are implemented for practical use can be biased or unrepresentative of real-world scenarios. Consequently, despite robust training, AI remains susceptible to errors. Misdiagnoses by AI have the potential to result in incorrect decisions and inadvertent harm to patients.

Who Bears Responsibility?

Determining responsibility in such scenarios presents a challenging inquiry: should the clinician using AI as a decision support tool shoulder the responsibility, or should it fall upon the software company? [41] While no straightforward answer exists, this query brings attention to the issue of liability, thereby delving into the concepts of responsible AI and the responsible utilization of AI in healthcare.

Responsible AI

Responsible AI implementation in healthcare serves as a guiding framework rather than a final product. It delineates how stakeholders in the health sector can address AI-related risks by fulfilling specific criteria. This framework is rooted in three crucial elements: accountability, responsibility, and transparency [42, 43] (Fig. 9.2).

- **Accountability**: In terms of accountability, the trust instilled in AI devices influences healthcare professionals and patients to develop convictions based on AI-driven decisions or predictions. Consequently, accountable AI should possess the ability to clarify, contextualize, and prioritize its decisions or predictions in accordance with moral and ethical considerations.
- **Responsibility**: The principle of responsibility spans the entire spectrum of participants, encompassing developers, manufacturers, suppliers, users, and the AI program itself. Each entity is obligated to take responsibility for its decisions, openly acknowledging limitations by exposing errors and unforeseen outcomes.

Notably, healthcare professionals hold responsibility for their decision-making in patient care, including how they interpret AI predictions and subsequently act on them. Therefore, comprehensive education and training on AI technology, such as the offerings from AIProHealth, become crucial for healthcare professionals. AIProHealth represents an international consortium where experts collaborate to create a customizable AI learning environment featuring diverse learning modules [44].

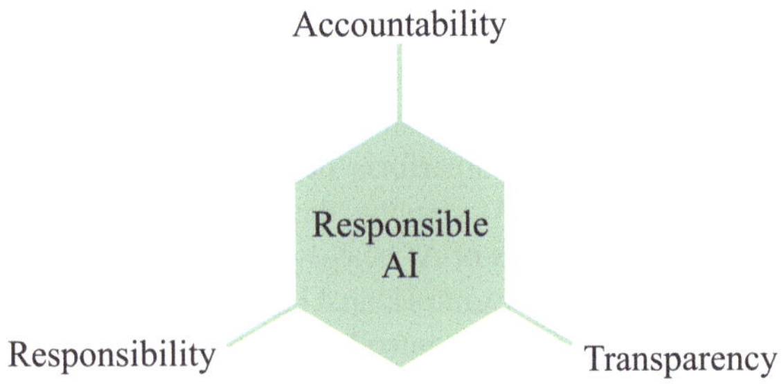

Fig. 9.2 Key elements for responsible AI development. See text for further explanation

- **Transparency**: The transparency pillar responds to the demands of regulators and users for clarity in responsible AI [45]. The origin and dynamics of input data must be transparent to guarantee fair collection, generation, and management, thereby minimizing biases. Additionally, AI decision-making mechanisms should be transparent and open for inspection, testing, and adjustments throughout their life cycle.

Adherence to these principles demands commitment from all stakeholders, prioritizing human values and societal welfare at the core of AI systems. This approach aims to ensure the responsible use of AI in healthcare, upholding ethical standards and mitigating potential risks linked to its implementation.

Practicing Responsible AI

Embracing AI responsibly within healthcare goes beyond its mere implementation; it necessitates the conscientious use of AI by all stakeholders. In March 2021, the American Medical Association outlined seven key insights aimed at fostering responsible AI utilization in healthcare [46] (Fig. 9.3):

1. **Advocacy for inclusive data practices**: Advocate for data that represents the population, emphasizing accessibility, standardization, and quality. Instead of a uniform approach, train AI to factor in individual needs. Ensure AI is precisely tailored for all demographics and meets standard criteria for quality assurance. Maintain consistent accessibility and transferability of AI data at all times.
2. **Prioritizing ethical and inclusive AI**: Give precedence to ethical, fair, and inclusive medical AI while confronting both explicit and implicit biases. Evaluate AI systems based on their capacity to tackle current biases and AI-induced prejudices that might worsen existing inequalities. Following this assessment, experts can determine whether to validate, invalidate, or use the system in restricted scenarios.
3. **Contextualizing transparency and trust**: Adjust the conversation about transparency and trust to suit a variety of needs. Understand that tailoring AI applications and transparency standards should consider the specific requirements of

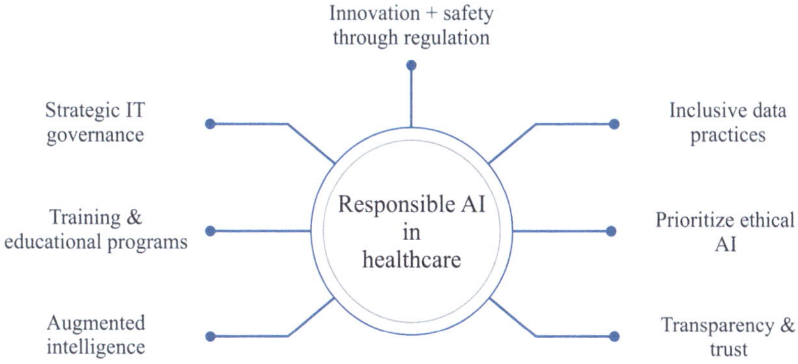

Fig. 9.3 Responsible AI in healthcare. See text for further explanation

users, the societal landscape, environmental factors, and technical intricacies. Although complete disclosure of AI training data is vital, total algorithmic transparency might not consistently yield positive outcomes. Establish precise guidelines for outlining data, performance, and algorithmic transparency.
4. **Emphasis on augmented intelligence**: Shift the focus to enhanced intelligence instead of entirely autonomous AI entities in the near term [29, 47]. Recognize the significant technical and regulatory hurdles that autonomous AI poses, grasping that prioritizing independent AI agents might hinder the efficient use of AI in healthcare. Enhancing the capabilities of doctors through augmented intelligence stands as a more feasible approach for now.
5. **Development of training and educational programs**: Develop and put into action thorough training and educational initiatives for every user, encompassing healthcare professionals and patients alike. Providing direction on managing clinical AI is essential to foster trust and secure informed consent from patients, as their willingness to embrace AI in healthcare relies heavily on their comprehension of medical AI.
6. **Strategic IT governance**: Establish a well-defined and knowledgeable information technology (IT) governance plan prior to incorporating AI into healthcare systems. A capable and resilient IT governance strategy is crucial for preparedness, mitigating potential challenges such as failures in AI integration and breaches in data protection.
7. **Balancing innovation with safety through regulation**: Find equilibrium between innovation and safety by enacting effective regulations and legislation. Evaluate clinical AI for its effectiveness and safety across its entire lifespan, implementing stringent control and enforcement measures to uphold user trust.

In adhering to these recommendations for responsible AI, numerous practical and ethical challenges come to the forefront, underscoring the complexity of integrating AI into healthcare responsibly.

In an effort to oversee AI practices in the healthcare domain within the United States, a recent move involved the issuance of an "executive order on the safe, secure, and trustworthy advancement of artificial intelligence" [48].

According to the directive issued in October 2023, the US federal government aims to champion responsible AI safety and security principles, collaborating with other nations, including competitors [48]. By taking a leadership role in global discussions and partnerships, the objective is to ensure that AI contributes to the well-being of the entire world, mitigating the risk of widening disparities, upholding human rights, and preventing potential harms.

References

1. Aristotle: Ethics and the Virtues, URL: https://www.austincc.edu/achristi/aristotle.html
2. Kant: The Moral Order, URL: https://www.austincc.edu/achristi/kant.html#:~:text=Constrained%20only%20by%20the%20principle,agent%20regards%20itself%20as%20determining%2C

3. The general data protection regulation, URL: https://www.consilium.europa.eu/en/policies/data-protection/data-protection-regulation/#:~:text=The%20GDPR%20lists%20the%20rights,his%20or%20her%20personal%20data
4. FDA, Proposed Regulatory Framework for Modifications to Artificial Intelligence/Machine Learning (AI/ML)-Based Software as a Medical Device (SaMD), URL: https://www.fda.gov/files/medical%20devices/published/US-FDA-Artificial-Intelligence-and-Machine-Learning-Discussion-Paper.pdf
5. FDA, Artificial Intelligence/Machine Learning (AI/ML)-Based Software as a Medical Device (SaMD) Action Plan, URL: https://www.fda.gov/media/145022/download
6. FDA, Premarket Approval (PMA), October 3, 2022, URL: https://www.fda.gov/medical-devices/premarket-submissions-selecting-and-preparing-correct-submission/premarket-approval-pma
7. Software as a Medical Device (SaMD): Key Definitions: http://www.imdrf.org/docs/imdrf/final/technical/imdrf-tech-131209-samd-keydefinitions-140901.pdf.
8. Software as a Medical Device (SaMD): Clinical Evaluation: https://www.fda.gov/downloads/medicaldevices/deviceregulationandguidance/guidancedocuments/ucm524904.pdf.
9. Deciding When to Submit a 510(k) for a Software Change to an Existing Device: https://www.fda.gov/downloads/medicaldevices/deviceregulationandguidance/guidancedocuments/ucm514737.pdf.
10. 21 CFR 807.81(a)(3). Modifications to a device approved through a PMA are governed by the criteria in 21 CFR 814.39(a). Modifications to Devices Subject to Premarket Approval (PMA) – The PMA Supplement Decision-Making Process: https://www.fda.gov/downloads/MedicalDevices/DeviceRegulationandGuidance/GuidanceDocuments/UCM089360.pdf.
11. WHO, WHO outlines considerations for regulation of artificial intelligence for health, 19 October 2023, URL: https://www.who.int/news/item/19-10-2023-who-outlines-considerations-for-regulation-of-artificial-intelligence-for-health
12. FDA, Artificial Intelligence and Machine Learning (AI/ML)-Enabled Medical Devices, URL: https://www.fda.gov/medical-devices/software-medical-device-samd/artificial-intelligence-and-machine-learning-aiml-enabled-medical-devices
13. Professor Paul Kinahan, PhD, FDA Publishes List Of AI-Enabled Medical Devices, UW Medicine, Department of Radiology, URL: https://rad.washington.edu/news/fda-publishes-list-of-ai-enabled-medical-devices/
14. Benjamens, S., Dhunnoo, P. & Meskó, B. The state of artificial intelligence-based FDA-approved medical devices and algorithms: an online database. npj Digit. Med. 3, 118 (2020). https://doi.org/10.1038/s41746-020-00324-0
15. FDA, Artificial Intelligence and Machine Learning in Software as a Medical Device, URL: https://www.fda.gov/medical-devices/software-medical-device-samd/artificial-intelligence-and-machine-learning-software-medical-device#:~:text=In%20this%20potential%20approach%2C%20the,part%20of%20the%20approved%20pre%2D
16. U.S. Department of Health and Human Services, Summary of the HIPAA Privacy Rule, URL: https://www.hhs.gov/hipaa/for-professionals/privacy/laws-regulations/index.html
17. The Regulation of Artificial Intelligence as a Medical Device, November 2022, URL: https://assets.publishing.service.gov.uk/media/6384bf98e90e0778a46ce99f/RHC_regulation_of_AI_as_a_Medical_Device_report.pdf
18. WHO, Ethics and governance of artificial intelligence for health, June 28, 2021, URL: https://www.who.int/publications/i/item/9789240029200
19. GPAI, URL: https://gpai.ai/
20. Ministry of Electronics & IT, GPAI 2023 begins in New Delhi from tomorrow, 11 DEC 2023, URL: https://pib.gov.in/PressReleaseIframePage.aspx?PRID=1985143#:~:text=The%20Global%20Partnership%20on%20AI,idea%20developed%20within%20the%20G7.
21. EU AI Act: first regulation on artificial intelligence, URL: https://www.europarl.europa.eu/topics/en/article/20230601STO93804/eu-ai-act-first-regulation-on-artificial-intelligence
22. Medicines and Healthcare products Regulatory Agency, Good Machine Learning Practice for Medical Device Development: Guiding Principles, 27 October 2021, URL: https://www.gov.

uk/government/publications/good-machine-learning-practice-for-medical-device-development-guiding-principles/good-machine-learning-practice-for-medical-device-development-guiding-principles
23. Medicines and Healthcare products Regulatory Agency, Predetermined Change Control Plans for Machine Learning-Enabled Medical Devices: Guiding Principles, 24 October 2023, URL: https://www.gov.uk/government/publications/predetermined-change-control-plans-for-machine-learning-enabled-medical-devices-guiding-principles
24. FDA, Digital Health Software Precertification (Pre-Cert) Pilot Program, URL: https://www.fda.gov/medical-devices/digital-health-center-excellence/digital-health-software-precertification-pre-cert-pilot-program
25. FDA selects participants for new digital health software precertification pilot program, URL: https://www.fda.gov/news-events/press-announcements/fda-selects-participants-new-digital-health-software-precertification-pilot-program
26. Spiekermann, S. IEEE P7000—The First Global Standard Process for Addressing Ethical Concerns in System Design. Proceedings 2017, 1, 159. https://doi.org/10.3390/IS4SI-2017-04084
27. ISO/IEC JTC 1/SC 42, Artificial intelligence, URL: https://www.iso.org/committee/6794475.html
28. The IEEE Global Initiative on Ethics of Autonomous and Intelligent Systems, URL: https://standards.ieee.org/industry-connections/ec/autonomous-systems/
29. Brady, A.P., Allen, B., Chong, J. et al. Developing, purchasing, implementing and monitoring AI tools in radiology: practical considerations. A multi-society statement from the ACR, CAR, ESR, RANZCR & RSNA. Insights Imaging 15, 16 (2024). https://doi.org/10.1186/s13244-023-01541-3
30. Abouelmehdi, K., Beni-Hessane, A. & Khaloufi, H. Big healthcare data: preserving security and privacy. J Big Data 5, 1 (2018). https://doi.org/10.1186/s40537-017-0110-7
31. Gupta, D.S., Mazumdar, N., Nag, A. et al. Secure data authentication and access control protocol for industrial healthcare system. J Ambient Intell Human Comput 14, 4853–4864 (2023). https://doi.org/10.1007/s12652-022-04370-2
32. Eichelberg, M., Kleber, K. & Kämmerer, M. Cybersecurity in PACS and Medical Imaging: an Overview. J Digit Imaging 33, 1527–1542 (2020). https://doi.org/10.1007/s10278-020-00393-3
33. Priyanka, Singh, A.K. A survey of image encryption for healthcare applications. Evol. Intel. 16, 801–818 (2023). https://doi.org/10.1007/s12065-021-00683-x
34. Ntantogian, C., Laoudias, C., Honrubia, A.J.D., Veroni, E., Xenakis, C. (2021). Cybersecurity Threats in the Healthcare Domain and Technical Solutions. In: Vlamos, P., Kotsireas, I.S., Tarnanas, I. (eds) Handbook of Computational Neurodegeneration. Springer, Cham. https://doi.org/10.1007/978-3-319-75479-6_38-1
35. MIDRC, URL: https://www.midrc.org/
36. Larobina M. Thirty Years of the DICOM Standard. Tomography. 2023 Oct 6;9(5):1829–1838. https://doi.org/10.3390/tomography9050145. PMID: 37888737; PMCID: PMC10610864.
37. Doel T, Shakir DI, Pratt R, Aertsen M, Moggridge J, Bellon E, David AL, Deprest J, Vercauteren T, Ourselin S. GIFT-Cloud: A data sharing and collaboration platform for medical imaging research. Comput Methods Programs Biomed. 2017 Feb;139:181–190. https://doi.org/10.1016/j.cmpb.2016.11.004. Epub 2016 Nov 10. PMID: 28187889; PMCID: PMC5312116.
38. ARPA-H BDF Toolbox, URL: https://arpa-h.gov/research-and-funding/programs/arpa-h-bdf-toolbox
39. U.S. Department of Health and Human Services, Summary of the HIPAA Security Rule, URL: https://www.hhs.gov/hipaa/for-professionals/security/laws-regulations/index.html#:~:text=General%20Rules&text=Ensure%20the%20confidentiality%2C%20integrity%2C%20and,impermissible%20uses%20or%20disclosures%3B%20and
40. Lax, G., Nardone, R. & Russo, A. Enabling secure health information sharing among healthcare organizations by public blockchain. Multimed Tools Appl (2024). https://doi.org/10.1007/s11042-024-18181-4

41. Neri, E., Coppola, F., Miele, V. et al. Artificial intelligence: Who is responsible for the diagnosis?. Radiol med 125, 517–521 (2020). https://doi.org/10.1007/s11547-020-01135-9
42. Trocin, C., Mikalef, P., Papamitsiou, Z. et al. Responsible AI for Digital Health: a Synthesis and a Research Agenda. Inf Syst Front 25, 2139–2157 (2023). https://doi.org/10.1007/s10796-021-10146-4
43. Dignum, V. (2023). Responsible Artificial Intelligence: Recommendations and Lessons Learned. In: Eke, D.O., Wakunuma, K., Akintoye, S. (eds) Responsible AI in Africa. Social and Cultural Studies of Robots and AI. Palgrave Macmillan, Cham. https://doi.org/10.1007/978-3-031-08215-3_9
44. AIProHealth, Practical Artificial Intelligence for Healthcare Professionals, URL: https://umcgresearch.org/w/dash-project-aiprohealth#:~:text=AIProHealth%20is%20a%20large%20international,aspect%20have%20already%20been%20developed.
45. Najjar, R. Redefining Radiology: A Review of Artificial Intelligence Integration in Medical Imaging. Diagnostics 2023, 13, 2760. https://doi.org/10.3390/diagnostics13172760
46. AMA, 7 tips for responsible use of health care AI, URL: https://www.ama-assn.org/practice-management/digital/7-tips-responsible-use-health-care-ai
47. Sorantin, E., Grasser, M.G., Hemmelmayr, A. et al. The augmented radiologist: artificial intelligence in the practice of radiology. Pediatr Radiol 52, 2074–2086 (2022). https://doi.org/10.1007/s00247-021-05177-7
48. The White House, Executive Order on the Safe, Secure, and Trustworthy Development and Use of Artificial Intelligence, https://www.whitehouse.gov/briefing-room/presidential-actions/2023/10/30/executive-order-on-the-safe-secure-and-trustworthy-development-and-use-of-artificial-intelligence/#:~:text=The%20Federal%20Government%20will%20seek,inequities%2C%20threatening%20human%20rights%2C%20and
49. Seh AH, Zarour M, Alenezi M, Sarkar AK, Agrawal A, Kumar R, Khan RA. Healthcare Data Breaches: Insights and Implications. Healthcare (Basel). 2020 May 13;8(2):133. https://doi.org/10.3390/healthcare8020133. PMID: 32414183; PMCID: PMC7349636.
50. The HIPAA Journal, 7 Ways AI Can be Used by Hackers to Steal Healthcare Data, Oct 13, 2023, URL: https://www.hipaajournal.com/ai-hackers-healthcare-data/#:~:text=However%2C%20AI%20can%20also%20be,receiving%20timely%20and%20effective%20treatment.

Chapter 10
Future Trends and Challenges

10.1 Introduction

Recent advancements in AI within medical imaging have ignited a substantial shift, reshaping diagnostic approaches and patient care. AI's advancements in this domain hold the promise of significantly enhancing precision in diagnoses and treatment planning, surpassing traditional methods [1].

Methodologies such as explainable AI and federated learning, coupled with innovative techniques such as multimodal imaging, AR, VR, and GANs, demonstrate potential through validated applications in healthcare settings.

The imminent integration of AI across clinical domains indicates a pivotal moment, where AI algorithms are positioned to excel in interpreting imaging data. This convergence of AI and clinical practice stands to redefine medical procedures, elevating standards in healthcare. However, achieving this significant change requires careful planning for seamless integration, navigating through various challenges.

10.2 Recent Advances in AI for Medical Imaging

Recent advances in artificial intelligence (AI) have ushered in a pivotal era within the realm of medical imaging, fundamentally reshaping the landscape of diagnosis, treatment procedures, and patient monitoring. The evolution of AI technologies has revolutionized the interpretation of complex medical images, exhibiting capabilities that often rival or surpass human expertise.

This shift is primarily a result of the progress in AI applied to medical imaging. Delving into these advancements reveals how AI is redefining medical imaging technology (Fig. 10.1).

E. Seeram, V. Kanade, *Artificial Intelligence in Medical Imaging Technology*,
https://doi.org/10.1007/978-3-031-64049-0_10

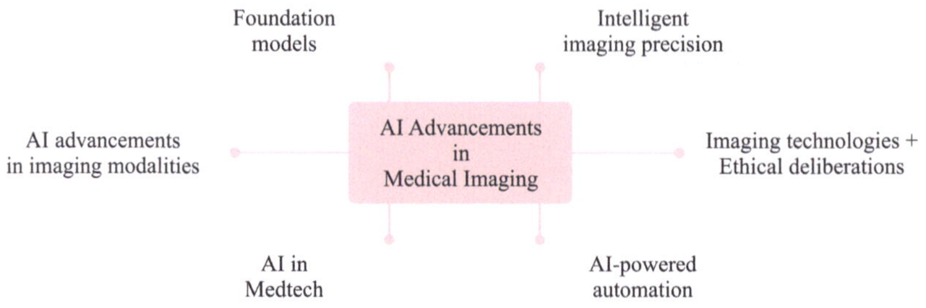

Fig. 10.1 AI advancements in medical imaging. See text for further explanation

1. **Intelligent Imaging Precision**

AI's impact on medical imaging has been profound, notably in significantly enhancing diagnostic accuracy. Notably, credible sources have highlighted how AI leverages machine learning algorithms to discern intricate patterns and anomalies in medical images, enabling the detection of diseases, including cancer, in their nascent stages with remarkable precision [2, 3]. These AI systems, trained on extensive datasets, excel in identifying subtle indicators that might elude human observation.

2. **AI Advancements in Imaging Modalities**

AI's integration into magnetic resonance imaging (MRI), positron emission tomography (PET), and computed tomography (CT) scans has witnessed substantial progress. According to News-Medical.net, specialized software driven by AI has bolstered image clarity by mitigating blurriness and enhancing diagnostic precision [4]. Moreover, AI algorithms adeptly reconstruct images even with reduced radiation doses or shorter scan times, ensuring diagnostic integrity while prioritizing patient safety.

Additionally, the integration of AI with imaging modalities such as ultrasound, X-ray, and nuclear medicine, as emphasized by the Radiological Society of North America (RSNA), signals a groundbreaking era for radiologists [5]. This integration equips radiologists with AI-powered tools to discern patterns indicative of complex diseases, designing more effective diagnostic approaches. The burgeoning interest in AI-assisted imaging, outlined by diagnostic imaging's monthly roundup, signifies the dynamic evolution of this field [6].

3. **The Emergence of Foundation Models**

A notable leap in AI for medical imaging lies in the advent of "foundation models," as highlighted by *Nature* magazine [7]. These adaptable AI models, trained on diverse datasets and tasks, offer versatility across various medical imaging applications. Their adaptability allows for seamless integration into diverse diagnostic workflows within different healthcare institutions, marking a significant stride toward consolidating AI's role in the medical imaging landscape.

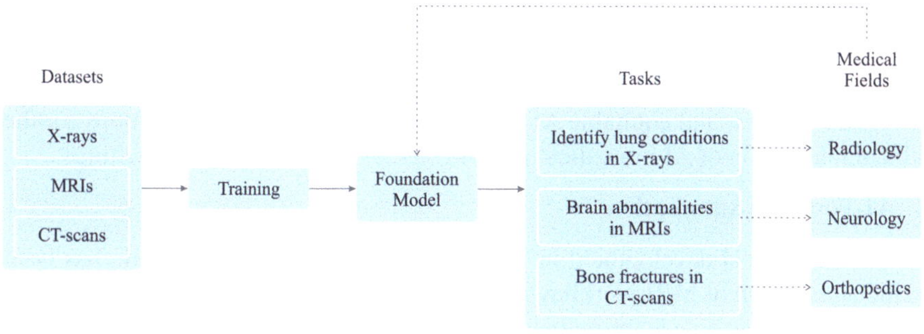

Fig. 10.2 Example of a foundation model. See text for further explanation

Imagine a foundation model developed specifically for medical imaging. This model undergoes training on an extensive array of datasets comprising X-rays, MRIs, CT scans, and other medical imaging data from various sources and specialties. Traditional AI models might be trained to excel in one specific imaging task, such as identifying anomalies in chest X-rays. However, a foundation model is designed to be more versatile [8]. It's trained on a multitude of tasks, not just focusing on a single type of imaging or a specific abnormality.

For instance, it learns to identify lung conditions in X-rays, brain abnormalities in MRIs, and bone fractures in CT scans, among other tasks as depicted in Fig. 10.2. This diverse training equips the foundation model with a broader understanding of different medical imaging domains.

Now, consider a hospital that has departments specializing in different medical fields such as radiology, neurology, and orthopedics. Each department traditionally uses separate AI systems tailored to their specific imaging needs. However, with the advent of foundation models, these departments can leverage the same adaptable AI model across various imaging tasks.

This ability to cater to diverse imaging needs without the need for substantial customization or retraining marks a significant advancement. It streamlines processes within healthcare institutions, allowing for a more standardized and effective application of AI in medical imaging, ultimately enhancing diagnostic precision and patient care.

4. **AI in Medtech**

AI's influence extends beyond medical imaging, permeating medtech spheres, as highlighted by Medical Design & Outsourcing [9]. The rise of generative AI technologies, exemplified by entities like OpenAI and Microsoft, holds promise in creating synthetic medical images for educational and research purposes, alleviating privacy concerns tied to patient data usage [10].

5. **Synergy of Imaging Technologies and Ethical Deliberations**

Advancements in imaging hardware, discussed at the European Congress of Radiology (ECR) Congress and reported by DirectIndustry e-Magazine, complement AI's progress [11]. These improvements in imaging speed, resolution, and functionality empower AI algorithms with superior input data, thereby amplifying diagnostic efficacy. Nonetheless, ethical deliberations on algorithmic transparency, data privacy, and equitable AI access persist as integral considerations for the ethical integration of AI in medical imaging across diverse healthcare settings.

6. **AI-Powered Automation**

Another pivotal development in 2023, documented by Potočnik et al., has been in the automation of MRI workflows [12]. TheHealthSite.com has also stressed the role of AI and machine learning in expediting radiology's progress through process automation [13].

Behold.ai, a London-based company, serves as a prime example of AI-powered automation in the medical field [14]. Their AI-driven medical diagnosis platform specializes in CT scans and X-rays, swiftly identifying conditions such as lung cancer and stroke within 30 seconds. Using AI-enabled autonomous diagnosis software, the platform examines chest X-rays, accurately annotating findings as heatmaps with over 90% accuracy. Collaborating with the NHS (National Health Services), Behold.ai has successfully implemented its AI solution across multiple NHS trusts [15].

Presently, AI-driven tools significantly enhance and standardize imaging processes, previously susceptible to human variations. They ensure consistent and reliable image capture, alongside improving post-processing procedures, including segmentation and quantification of specific tissues.

Key Developments

AI progress has completely changed the face of medical imaging, empowering clinicians in diagnosing, planning therapies, and tracking patient progress. The rising integration of AI in medical imaging is evident in BioSpace's August 2023 projection for the global AI in medical imaging market, indicating significant financial expansion. The leap from about $762.84 million in 2022 to an estimated $14,423.15 million by the next decade highlights the rising investment in AI technologies [16]. This investment holds the promise of improving healthcare results and operational effectiveness.

Medical imaging, a crucial aspect of modern healthcare, has benefited significantly from the infusion of AI, particularly with the development of machine learning and deep learning techniques. Deep learning, especially convolutional neural networks (CNNs), stands at the forefront of these AI-driven advances. CNNs are known for their prowess in image classification, object detection, and segmentation tasks, which are foundational for medical image analysis.

This expertise forms the basis for AI applications in radiology, pathology, dermatology, ophthalmology, and various other specialties that rely heavily on imaging. The following section explores recent instances where AI has sparked significant changes in different medical imaging domains.

1. **Radiology**

   AI algorithms in radiology have showcased remarkable efficacy, evident in studies published in esteemed medical journals such as the *Lancet* and *Nature Medicine* [63]. Examples include Stanford University's research published in November 2017, highlighting deep learning models' prowess in accurately detecting pneumonia in chest X-rays, rivaling expert radiologists [17]. These advancements pave the way for AI-driven tools facilitating swift diagnosis, crucial for timely interventions and improved patient outcomes.

2. **Ophthalmology and Retinal Imaging**

   Collaborations between global tech giants such as Google, IBM, and leading healthcare institutions have propelled AI advancements in interpreting retinal images. Google's DeepMind collaboration with Moorfields Eye Hospital yielded AI models capable of early diabetic retinopathy and macular degeneration detection [18]. Published in *Nature Medicine*, these findings emphasize the potential for precise diagnosis, mitigating vision loss in patients.

3. **Pathology**

   The integration of AI into pathology, notably in partnerships with Massachusetts General Hospital and MIT, has revolutionized cancer diagnostics [19]. Research published in "National Library of Medicine" showcases AI-powered systems aiding pathologists in identifying and grading cancerous tissue with unparalleled accuracy [20]. Companies such as PathAI leverage these findings to develop AI-assisted pathology platforms, enhancing diagnostic precision and enabling tailored treatment strategies.

4. **Cardiology**

   Collaborations between tech companies and leading cardiology centers, exemplified by GE Healthcare, have yielded AI-powered solutions in interpreting echocardiograms and detecting cardiac abnormalities [21]. Clinical trials corroborate the efficacy of these AI-driven tools in early detection and treatment optimization for cardiovascular diseases.

5. **Neurology**

   In neurology, AI's implementation in stroke diagnosis has been game-changing [22]. Collaborative efforts between research institutions and AI startups have led to algorithms capable of rapidly analyzing brain imaging scans. This innovation aids in prompt interventions, significantly reducing morbidity and mortality associated with cerebrovascular incidents.

6. **Oncology**

   The convergence of AI and medical imaging in oncology has redefined cancer diagnosis and personalized treatment. Collaborations such as IBM Watson Health with leading cancer centers yield AI systems analyzing extensive datasets [23].

Studies published in the *Journal of Clinical Oncology* emphasize AI-driven decision support systems in tailoring treatment plans and enhancing therapeutic outcomes and patient survival rates [24]. Moreover, Zhixiang Wang's doctoral thesis, published by Maastricht University, comprehensively explores various AI applications in the field of oncology [25].

7. **AI Accessibility in Underserved Regions**

AI's integration with imaging technologies has expanded healthcare accessibility, particularly in resource-constrained settings. Supported by RAD-AID and tech partnerships, portable imaging devices embedded with AI algorithms offer accurate diagnostics and improved patient care in underserved regions [26]. Field trials and published data validate these solutions' effectiveness.

The mentioned examples underscore the importance of interdisciplinary collaboration across various fields. The fusion of new technologies with medical expertise is profoundly changing how healthcare is provided. Together, clinicians, data scientists, and AI experts are turning technological advancements into practical advantages for patients.

10.3 Emerging Technologies (e.g., Explainable AI, Federated Learning)

New technologies in medical imaging, such as explainable AI and federated learning, demonstrate innovative approaches that are continually bringing advancements to healthcare. These advancements leverage novel techniques such as multimodal imaging, augmented reality (AR), virtual reality (VR), and generative adversarial networks (GANs) [27–29]. They present promising applications validated by research papers, collaborations between global tech companies and medical facilities, and implementations within healthcare institutions.

The following sections delves deeper into each upcoming advancement for a comprehensive understanding (Fig. 10.3).

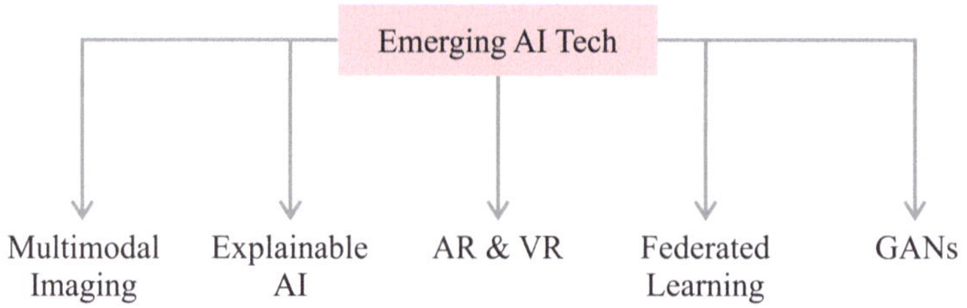

Fig. 10.3 Emerging AI tech. See text for further explanation

1. **Explainable AI**

Explainable AI (XAI) in medical imaging is a field that aims to make AI systems more transparent and interpretable, ensuring that healthcare professionals can understand and trust the decisions made by these systems. One notable application is in radiology, where AI algorithms often operate as black boxes, making it challenging for clinicians to comprehend the reasoning behind a specific diagnosis.

Recent research papers and studies, such as those published in the *Journal of the American Medical Association* (*JAMA*) and NIH, have explored the integration of explainable AI in radiological applications [30, 31]. For instance, algorithms that can highlight the regions of interest in medical images and provide a rationale for their decisions offer clinicians valuable insights into the diagnostic process. This not only aids in building trust in AI-assisted diagnoses but also allows radiologists to validate and potentially improve the accuracy of the results.

Global tech companies, recognizing the importance of explainability in AI, have started incorporating XAI principles into their medical imaging solutions. IBM's Watson Health, for example, has been working on AI models for medical imaging that provide not only accurate predictions but also clear explanations of how the model arrived at a particular conclusion [32]. This transparency is crucial for gaining regulatory approval, fostering acceptance among healthcare professionals, and facilitating the integration of AI into clinical workflows.

2. **Federated Learning**

Federated learning is another pivotal technology that addresses challenges related to data privacy and security in the context of medical imaging [33–35]. In traditional machine learning models, data is centralized, posing concerns about the privacy of sensitive patient information. Federated learning decentralizes this process, allowing models to be trained across multiple devices, institutions, or partners (i.e. hospitals, pharma companies, clinical research centers, medical imaging facilities, or individual researchers) without sharing raw data, as shown in Fig. 10.4.

Studies published in leading journals, such as *Nature* and *IEEE Transactions on Medical Imaging*, have explored the potential of federated learning in medical image analysis [36, 37]. For instance, collaborative efforts between institutions have demonstrated how models for diagnosing diseases, such as cancer or neurological disorders, can be trained across different healthcare centers without sharing patient-specific data. This not only protects patient privacy but also enables the development of robust and generalized models that perform well across diverse populations.

Healthcare institutions such as Mayo Clinic have also embraced federated learning to leverage collective intelligence while safeguarding patient privacy [38]. Partnerships between academic medical centers (such as Oxford Academic, National Institutes of Health, Springer, ACM Digital Library, and *Nature* journal) and technology providers have led to the implementation of federated learning frameworks in large-scale medical imaging studies [39]. This collaborative approach facilitates the development of AI models that are more representative of diverse patient

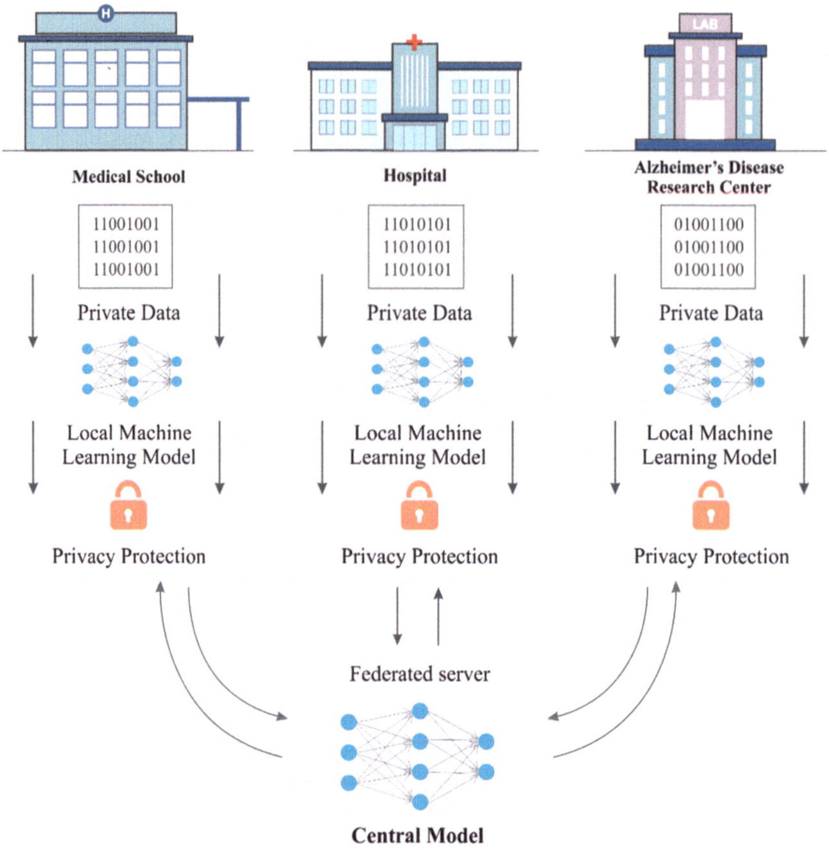

Fig. 10.4 An example of federated learning architecture for healthcare institutions. See text for further explanation

populations, ultimately improving the generalizability of these models in real-world clinical settings.

One compelling use case of federated learning is in pathology. PathAI, a company specializing in AI-powered pathology solutions, has employed federated learning techniques to train models across a network of pathology labs.

3. **Multimodal Imaging**

Multimodal imaging, an approach that combines different imaging modalities, has garnered attention for its potential to provide comprehensive and complementary information for more accurate diagnoses. Research published in prominent journals such as *Comput Math Methods Med.* demonstrates the effectiveness of multimodal imaging fusion in fields such as oncology and neuroimaging [40].

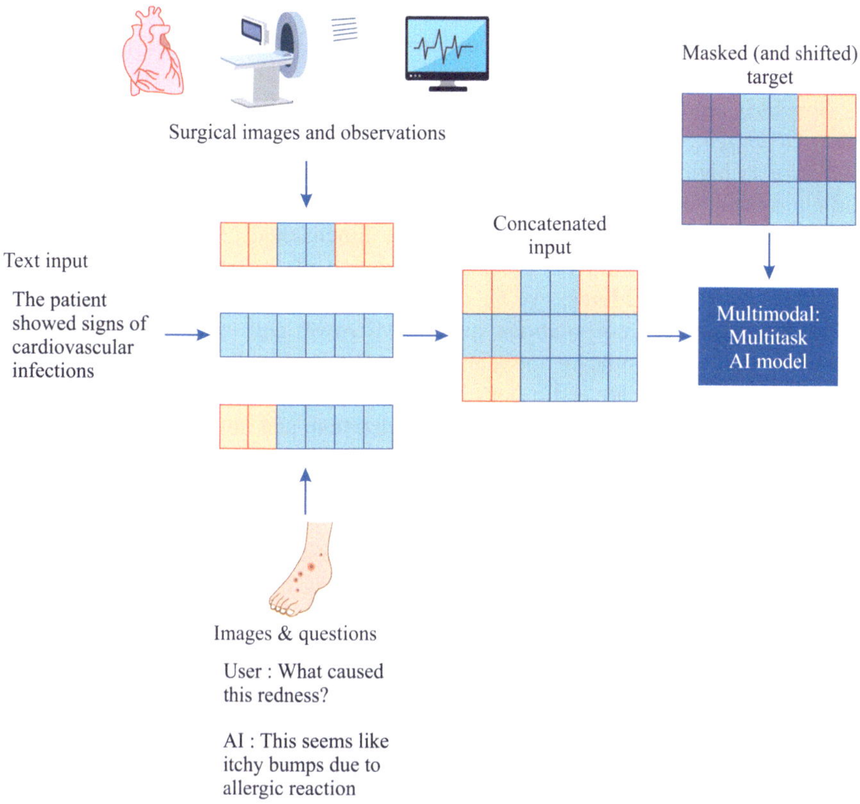

Fig. 10.5 Example of multimodal AI [27]. See text for further explanation

For instance, studies utilizing PET fused with MRI have shown enhanced precision in tumor delineation and staging for cancer patients. By combining PET's metabolic information with MRI's superior anatomical detail, clinicians can better characterize tumors and tailor treatment plans accordingly. Additionally, research in neuroimaging has explored the fusion of functional MRI (fMRI) with other modalities, enabling a deeper understanding of brain function and connectivity in conditions such as Alzheimer's disease and epilepsy [41]. An example of multimodal AI is depicted in Fig. 10.5.

Tech enterprises such as Siemens Healthineers and Philips have been at the forefront of multimodal imaging innovation [42–44]. Their integrated imaging systems and software solutions enable seamless fusion and visualization of data from various modalities, empowering clinicians with comprehensive insights for accurate diagnoses and treatment monitoring.

4. **Augmented Reality (AR) and Virtual Reality (VR)**

Augmented reality (AR) and virtual reality (VR) are other emerging technologies that hold immense potential in medical imaging. By overlaying digital information onto the real-world environment, AR facilitates enhanced visualization and navigation during surgical procedures, medical training, and preoperative planning.

Scholarly articles in prestigious scientific journals such as *IEEE Transactions on Medical Imaging* and the *International Journal of Computer Assisted Radiology and Surgery* highlight the utility of AR in guiding minimally invasive surgeries and aiding in complex interventions [45, 46]. For instance, AR-assisted navigation systems provide surgeons with real-time, three-dimensional (3D) visualizations of patient anatomy, allowing for precise localization of tumors, organs, or critical structures during procedures.

Multinational tech corporations such as Microsoft and its HoloLens platform have ventured into healthcare applications of AR/VR [47]. Collaborations between tech giants and healthcare institutions have resulted in the development of AR solutions for medical education, anatomical visualization, and intraoperative guidance, offering immersive experiences that enhance surgical precision and safety.

5. **Generative Adversarial Networks (GANs)**

As explained in earlier chapters, generative adversarial networks (GANs) represent a breakthrough in medical imaging by enabling the generation of synthetic, high-fidelity images. GANs consist of two neural networks, a generator and a discriminator, working in tandem to produce realistic images that mimic real patient data.

Scientific publications in journals such as *Nature Communications* and *Journal of Imaging* showcase GANs' capabilities in generating synthetic medical images with high resolution and anatomical accuracy [48, 49]. These synthetic images serve various purposes, including data augmentation for training AI models, enhancing limited datasets, and preserving patient privacy by generating anonymized images for research purposes.

Healthcare institutions and research collaborations have employed GANs to address challenges in medical imaging, particularly in scenarios where data scarcity or privacy concerns exist. Additionally, companies such as NVIDIA have developed GAN-based solutions tailored for medical imaging [50, 51]. Their platforms enable researchers and healthcare providers to generate synthetic images, simulate different imaging conditions, and improve the robustness of AI models by diversifying training datasets.

6. **Natural Language Processing (NLP)**

Natural language processing (NLP) is a branch of artificial intelligence that focuses on enabling computers to understand, interpret, and generate human language in a way that is meaningful and useful.

Today, companies such as "CloudMedX Health," are already utilizing NLP and deep learning to extract existing data from electronic medical records [52]. They provide clinical insights for healthcare professionals, aiding in the improvement of patient outcomes. While NLP plays a significant role in transcribing medical records and facilitating doctor-patient interactions, its potential benefits for the medical imaging field are even more pronounced in several ways:

- **Integration of clinical notes with imaging data**: NLP can analyze and extract relevant information from clinical notes, such as radiology reports or physician's notes, and correlate this information with medical images. By URLing textual information with imaging data, it provides a more comprehensive view of a patient's condition, aiding radiologists in making more informed diagnoses and treatment decisions.
- **Improved search and retrieval of imaging studies**: NLP can enhance the searchability of medical imaging databases by understanding natural language queries. Radiologists or clinicians can use conversational language to search for specific imaging studies, patient information, or specific findings within images, making the retrieval process faster and more efficient.
- **Automated report generation**: NLP algorithms can assist in generating structured and detailed reports based on the analysis of imaging data. By interpreting the findings from medical images and integrating this information with clinical knowledge, NLP can assist radiologists in creating comprehensive reports, saving time and ensuring consistency in reporting [64].
- **Clinical decision support**: NLP-powered systems can analyze large volumes of imaging and textual data to identify patterns, trends, or correlations that might not be immediately evident to human observers. This analysis can aid in predictive modeling, risk assessment, or identifying potential treatment responses, supporting radiologists and clinicians in making more accurate and personalized treatment plans.
- **Semantic understanding for image annotation**: NLP can assist in annotating medical images by understanding the semantic context of imaging reports or notes. This capability helps in automatically labeling or annotating images with relevant information, improving data organization and retrieval for future reference.

By leveraging NLP's capabilities, the medical imaging field can enhance the efficiency of radiologists and facilitate effective utilization of imaging technologies.

Despite these advancements, challenges persist in the adoption and optimization of these emerging technologies in medical field. Standardization, validation of synthetic images' clinical utility, and integration into clinical workflows remain areas of active research and development.

10.4 Potential Impact and Challenges in Adopting AI in Clinical Practice (e.g., Radiology)

The anticipated integration of AI across clinical practices, such as radiology, oncology, pathology, among others, is expected to profoundly influence medical diagnosis and patient care procedures for an extended period. AI algorithms can interpret imaging results with remarkable speed and accuracy, often surpassing human capability in repetitive or data-intensive tasks.

The following section explores how AI could impact its integration into clinical practice.

Potential Impact

The integration of AI into clinical practice could significantly change how the medical field operates, affecting it in various profound ways.

1. **Precision Diagnostics**

AI's ability to analyze vast amounts of medical data with speed and accuracy holds immense promise for precision diagnostics. Recent studies published in journals, such as *Nature Medicine*, showcase AI's effectiveness in interpreting medical images, detecting anomalies, and assisting in early disease diagnosis.

For instance, AI algorithms in radiology have demonstrated proficiency in identifying subtle abnormalities in X-rays and MRI scans, often surpassing human capabilities in terms of accuracy and speed. In certain arenas, such as brain imaging, where rapid detection is crucial, AI can provide near-instantaneous analysis that is critical for accessing treatment quickly.

Several institutions are leading the way in incorporating AI into radiology. For instance, Radiology Partners has launched an AI integration platform with AWS HealthImaging, aimed at enhancing diagnostic capabilities across its network [53]. Moreover, major health systems, including Mass General Brigham, have adopted advanced AI copilot capabilities from Nuance, enhancing precision and workflow in radiology departments [54].

2. **Workflow Optimization**

AI has the potential to streamline and enhance various aspects of healthcare operations. Administrative tasks, such as scheduling, billing, pre-fill reports with templated findings, and simple data entry, can be automated, allowing healthcare professionals to focus more on complex interpretation and decision-making tasks. Moreover, it can sort and prioritize imaging studies based on urgency or complexity, ensuring that critical cases are reviewed promptly.

Additionally, AI-powered predictive analytics can optimize hospital workflows by predicting patient admission rates, resource utilization, and disease outbreaks, facilitating proactive and efficient resource allocation.

3. **Personalized Treatment Plans**

The integration of AI allows for the development of personalized treatment plans based on individual patient data. By analyzing patient records, genetic information, and treatment outcomes, AI can assist healthcare providers in tailoring interventions to the specific needs of each patient. This personalized approach has been exemplified in oncology, where AI-driven models help predict treatment responses and recommend personalized cancer therapies.

4. **Patient Volume Management**

With the efficiency that AI brings, healthcare institutions can handle larger volumes of imaging studies without compromising the quality of care. This capacity expansion is particularly significant as the demand for medical imaging grows with a global aging population and increased screening programs.

5. **Remote Monitoring and Telemedicine**

The advent of AI-powered remote monitoring tools enhances the ability to manage and monitor patients outside traditional healthcare settings. Wearable devices (such as Fitbit Sense, Garmin Vivosmart 4, BioBeat's medical-grade devices, AliveCor KardiaBand) equipped with AI algorithms can continuously track vital signs, detect anomalies, and provide real-time health insights [55]. This is particularly valuable for chronic disease management, allowing healthcare providers to intervene promptly based on AI-generated alerts, potentially reducing hospitalizations and improving patient outcomes.

Challenges in Adopting AI in Clinical Practice
Despite its potential, the adoption of AI in clinical practice confronts several challenges that healthcare organizations must overcome.

1. **Ethical Considerations**

The deployment of AI requires the processing of massive amounts of sensitive patient data, which raises concerns about data privacy and security. Ensuring AI systems are compliant with regulations such as the Health Insurance Portability and Accountability Act (HIPAA) and the General Data Protection Regulation (GDPR) is essential [70, 71].

Thus, the ethical implications of AI in healthcare demand careful attention. Issues related to patient privacy, consent, and the responsible use of AI-generated insights are critical. Recent high-profile cases of data breaches and privacy concerns highlight the need for robust ethical frameworks [56, 57]. Regulatory bodies worldwide, such as the US Food and Drug Administration (FDA) and the European Medicines Agency (EMA), are actively working on establishing guidelines to ensure the ethical deployment of AI in clinical arenas.

2. **Validation and Standardization**

Validating and standardizing AI algorithms for clinical use is a complex process. Ensuring that these algorithms perform reliably across diverse patient populations and healthcare settings is crucial for their acceptance. The lack of standardized validation processes poses challenges in gaining regulatory approval for AI-driven medical devices. Initiatives by regulatory agencies and collaborations between researchers and industry stakeholders are essential for establishing rigorous validation protocols.

3. **Integration with Existing Systems**

The significant challenge arises due to the heterogeneity of digital systems in healthcare, hindering the seamless integration of AI into existing healthcare systems. Many healthcare institutions often work with legacy systems or operate on diverse and sometimes outdated technology infrastructures. Such systems may not easily interface with modern AI applications, necessitating costly updates or even complete overhauls.

Hence, ensuring compatibility, interoperability, and a smooth transition without disrupting daily operations requires substantial planning and investment. Collaborations between technology developers and healthcare providers are essential to navigate these integration challenges successfully.

4. **Physician Training and Acceptance**

Physician acceptance and proficiency in using AI tools are paramount for successful integration into clinical practice. Recent surveys, including those published in medical journals such as *JAMA*, reveal varying levels of acceptance and awareness among healthcare professionals [58]. Addressing the educational needs of healthcare professionals and fostering a culture of AI adoption through training programs and collaborative initiatives are crucial steps toward overcoming resistance and ensuring effective utilization.

5. **Bias and Transparency**

AI algorithms can inadvertently perpetuate biases present in the training data, leading to disparities in healthcare outcomes. Recognizing and mitigating biases in AI models is imperative to ensure equitable healthcare delivery. Moreover, the lack of transparency in AI decision-making poses challenges in gaining trust from both healthcare professionals and patients. Ongoing research and initiatives focusing on bias detection, transparency, and explainability are essential for advancing the ethical use of AI in clinical practice.

6. **Interaction Between Humans and AI**

In addition to technical performance and workflow integration, the significance of intangible human factors in the development and implementation of AI tools for radiology cannot be overstated. While AI has made remarkable strides, achieving diagnostic performance comparable to human readers in certain scenarios, various factors can affect the diagnostic accuracy of AI tools in clinical settings.

The combination of human readers and AI tools has been proposed to enhance overall diagnostic accuracy by allowing humans to catch errors made by AI or vice versa. However, recent studies have raised doubts about this assumption, emphasizing the need to explore psychological phenomena that may bias decision-making during human-AI interaction. Automation bias, wherein humans excessively rely on automated systems such as AI-powered decision support tools, can undermine their oversight abilities. Research in mammography revealed that even experienced readers demonstrated this bias when an AI system suggested an incorrect BI-RADS category, leading to significantly compromised performance [65].

Conversely, algorithmic aversion, the tendency to dismiss information solely because it originates from AI, has also been observed. Radiologists and other physicians rated AI-generated information as less reliable compared to human expertise, indicating a potential reluctance to trust AI outputs.

Furthermore, human-AI interaction can be influenced by the design of the user interface (UI) [59]. While many radiologists preferred image overlays for pulmonary nodule detection, studies found that this UI configuration did not improve performance. Instead, a minimalist text-only UI was more effective. Similarly, in endoscopy, the use of a computer-aided system for polyp detection resulted in reduced eye movements and increased misinterpretation of normal mucosa, highlighting the need for careful UI design [66].

Addressing these issues requires increased awareness and education among users and stakeholders to ensure the safe integration of AI into clinical practice. Further research is essential to provide evidence-based guidelines for optimizing human-AI interaction and designing effective UIs that enhance rather than hinder diagnostic performance.

7. **Cost of Implementation**

Adopting AI technologies can be an expensive venture, with costs associated with purchasing the technology, integrating it into existing systems, and training personnel. Smaller practices may find the cost prohibitive, leading to a digital divide in healthcare provision.

In essence, the intrusion of AI into clinical practice heralds a future of advanced precision medicine, improved patient outcomes, and enhanced healthcare delivery. While adoption does face hurdles, from ensuring ethical use of AI to addressing financial and structural challenges, the potential benefits make overcoming these challenges a worthy endeavor. Continued collaboration among healthcare professionals, AI developers, policy makers, and regulatory bodies is critical to navigate these challenges and harness the full power of AI in healthcare facility.

10.5 Future AI: Considerations for AI Tool Development

When creating new AI tools for clinical use, developers should keep in mind several important factors. Here are some key points they should focus on:

1. **Practical Value of New Products**

Making sure new products in radiology AI are genuinely useful is crucial for fitting them into how things already work. Developers need to make sure their solutions make it easier to spot problems, outline areas of interest, make diagnoses, or predict what might happen in a patient's case. One big mistake is making technologies based only on the information available, rather than what doctors actually need. Tools such as the ACR DSI Define-AI directory can help developers make sure their tech meets real clinical needs [60, 61]. Getting doctors involved early in the process is really important too. It helps figure out if the ideas are doable and useful before spending a lot of time and money.

2. **Advantages over Current Clinical/Radiology Tools**

Proving that new AI tools are better than what's already out there is tough, especially if the developers aren't doctors themselves. It's easier when there are competitions with clear rules and rankings, such as the ones at the RSNA and MICCAI conferences. If there's no competition data available, developers need to compare their AI's performance with what existing software or doctors can do. Tests where multiple doctors read images can help show if the AI is improving diagnoses, but these can be complicated and expensive.

3. **Radiomics: Clarity and Transparency**

Radiomics, which involves getting lots of data from medical images, can be hard to understand. Many of the details aren't easy to see or explain, which makes it tough to connect them with what doctors see. To help with this, there are tools such as heatmaps and saliency maps that show which areas the AI is focusing on. Some researchers are even trying to make the AI explain its decisions in plain language. Learning from past research is important here, and if the AI is too hard to understand, it might need more testing to prove it works in different situations, like in real hospitals.

In conclusion, the development of AI tools for clinical use in radiology demands careful consideration of practical value, superiority over existing tools, and clarity in radiomics. Ensuring genuine utility, engaging clinicians early, and validating improvements through rigorous testing are essential. Incorporating feedback from real-world clinical settings and striving for transparency in AI decision-making will foster the creation of impactful and trustworthy tools for medical imaging.

Factors to Consider Long-Term Stability and Safety of AI Tools

It's crucial to keep a close eye on how AI models perform in real-world clinical settings to ensure they're safe and effective. The US FDA emphasizes this in its approach to regulating software as a medical device (SaMD), including AI used in imaging. Over time, the performance of AI models—both "locked" (unchanging) ones and potentially adaptive ones—can decline due to various factors such as changes in input data, updates to imaging equipment or software, or shifts in patient demographics [67, 68].

To address this, institutions should set up systems to continuously monitor AI performance as part of their local governance processes. Detecting any decline in performance early is crucial to prevent risks to patient safety and ensure proper care.

1. **Monitoring Data Drift Impacting Model Performance**

Monitoring data drift impacting model performance is crucial for ensuring the reliability of AI systems in medical imaging. For instance, in an MRI department, an AI algorithm designed to detect brain tumors may start producing inaccurate results due to changes in equipment calibration or shifts in patient demographics.

By continuously collecting and analyzing real-time data, healthcare institutions can detect such drifts.

For example, if the AI algorithm starts misidentifying tumors that were previously detected accurately, radiologists can cross-reference the AI outputs with their own findings to identify any imaging artifacts or errors. This ongoing monitoring allows institutions to adjust the AI algorithm or address underlying issues promptly, ensuring consistent and accurate performance over time.

Hence, to address issues such as data drift, which can affect AI performance, institutions can establish baseline characteristics for input data and monitor for any deviations from that baseline. If significant changes occur, it may be necessary to re-evaluate the AI model's performance and take corrective steps accordingly.

2. **Future Adaptation of Local Learning AI Algorithms**

Today, all AI tools that have obtained regulatory approval are static, meaning they are fixed and cannot be adjusted or improved locally. This limitation prevents healthcare institutions from fine-tuning these tools or incorporating adaptive learning techniques, which could potentially enhance their performance and adaptability.

However, there is a promising development with the recent release of a draft guidance by the US Food and Drug Administration (FDA). This guidance proposes a "predetermined change control plan" that could revolutionize how AI algorithms are managed post-approval [69]. Essentially, this plan would enable future modifications to commercial algorithms, allowing for both local tuning and continuous learning.

This means that healthcare institutions may have the flexibility to adapt and improve AI tools over time, tailoring them to specific local needs and incorporating new knowledge gained from ongoing data analysis and learning. This approach could lead to more dynamic and effective AI solutions in medical imaging and other healthcare applications, ultimately benefiting patient care. However, it's essential to ensure that any modifications made under this plan are rigorously evaluated to maintain safety and efficacy standards [67–69].

3. **Other Variables: Managing AI in Clinical Settings**

In the complex world of using AI in healthcare, it's crucial for provider organizations to establish formal oversight processes, known as AI governance. This helps manage the entire life cycle of AI products in clinical settings. It involves everything from acquiring effective AI tools to monitoring their performance over time. It also includes making adjustments to the local environment, such as scanner protocols and workflow integration, as needed.

One often overlooked aspect is the impact of staff turnover. As new people join the team, they need to be properly trained in using AI tools and adapting to new workflows. This responsibility falls on both the vendors providing the AI technology and the leaders within the healthcare institution.

In essence, effective AI governance ensures that everyone involved stays up-to-date and proficient in using AI to improve patient care [62]. It's a collaborative effort between technology providers and healthcare leaders to ensure smooth integration and ongoing effectiveness of AI in clinical practice.

References

1. TheHealthSite.com, URL: https://www.thehealthsite.com/diseases-conditions/technological-advances-in-medicine-how-ai-and-machine-learning-are-revolutionizing-medical-sector-1002255/
2. Koh, DM., Papanikolaou, N., Bick, U. et al. Artificial intelligence and machine learning in cancer imaging. Commun Med 2, 133 (2022). https://doi.org/10.1038/s43856-022-00199-0
3. Pinto-Coelho L. How Artificial Intelligence Is Shaping Medical Imaging Technology: A Survey of Innovations and Applications. Bioengineering (Basel). 2023 Dec 18;10(12):1435. https://doi.org/10.3390/bioengineering10121435. PMID: 38136026; PMCID: PMC10740686.
4. Danielle Ellis, B.Sc., AI-powered software can provide MRI-quality brain images from CT scans, https://www.news-medical.net/news/20231017/AI-powered-software-can-provide-MRI-quality-brain-images-from-CT-scans.aspx, October 17, 2023
5. Krishna Juluru, Hao-Hsin Shih, Krishna Nand Keshava Murthy, Pierre Elnajjar, Amin El-Rowmeim, Christopher Roth, Brad Genereaux, Josef Fox, Eliot Siegel, Daniel L. Rubin, Integrating AI Algorithms into the Clinical Workflow, Radiology: Artificial Intelligence Vol. 3, No. 6, Aug 4 2021, https://pubs.rsna.org/doi/full/10.1148/ryai.2021210013
6. Advances in AI – January 2024, January 26, 2024, Diagnostic Imaging Staff, URL: https://www.diagnosticimaging.com/view/advances-in-ai-january-2024
7. Zhou, Y., Chia, M.A., Wagner, S.K. et al. A foundation model for generalizable disease detection from retinal images. Nature 622, 156–163 (2023). https://doi.org/10.1038/s41586-023-06555-x
8. Moor, M., Banerjee, O., Abad, Z.S.H. et al. Foundation models for generalist medical artificial intelligence. Nature 616, 259–265 (2023). https://doi.org/10.1038/s41586-023-05881-4
9. Chris Newmarker, AI breakthroughs in medtech: 7 ways to enhance healthcare, URL: https://www.medicaldesignandoutsourcing.com/ai-artificial-intelligence-breakthroughs-medtech-enhance-healthcare/, May 30, 2023.
10. Research at Microsoft 2023: A year of groundbreaking AI advances and discoveries, URL: https://www.microsoft.com/en-us/research/blog/research-at-microsoft-2023-a-year-of-groundbreaking-ai-advances-and-discoveries/, December 22, 2023.
11. Abigail Saltmarsh, Advances and Challenges in Medical Imaging, URL: https://emag.medicalexpo.com/ecr-2023-advances-and-challenges-in-medical-imaging/, August 16, 2023.
12. Jaka Potočnik, Shane Foley, Edel Thomas, Current and potential applications of artificial intelligence in medical imaging practice: A narrative review, Journal of Medical Imaging and Radiation Sciences, Volume 54, Issue 2, 2023, Pages 376–385, ISSN 1939-8654, https://doi.org/10.1016/j.jmir.2023.03.033, (https://www.sciencedirect.com/science/article/pii/S1939865423000796)
13. Tavishi Dogra, Radiology Revolution: Role Of Automation In Shaping The Future Of Medical Imaging, URL: https://www.thehealthsite.com/diseases-conditions/radiology-revolution-role-of-automation-in-shaping-the-future-of-medical-imaging-1029414/, November 8, 2023.
14. Behold.ai, URL: https://www.behold.ai/
15. Dartford and Gravesham NHS Trust implements AI in radiology, URL: https://www.behold.ai/news/dartford-and-gravesham-nhs-trust-implements-ai-in-radiology/, 3 June 2019.
16. BioSpace, AI in Medical Imaging Market will Grow to $14,423.15 Million Over Next 10 Years, URL: https://www.biospace.com/article/releases/ai-in-medical-imaging-market-will-grow-to-14-423-15-million-over-next-10-years/#:~:text=Insights%20(168)-,AI%20in%20Medical%20Imaging%20Market%20will%20Grow,Million%20Over%20Next%2010%20Years&text=The%20global%20AI%20in%20medical,34.8%25%20from%202023%20to%202032, August 31, 2023.
17. Taylor Kubota, Stanford algorithm can diagnose pneumonia better than radiologists, URL: https://news.stanford.edu/2017/11/15/algorithm-outperforms-radiologists-diagnosing-pneumonia/, November 15, 2017.

18. De Fauw, J., Ledsam, J.R., Romera-Paredes, B. et al. Clinically applicable deep learning for diagnosis and referral in retinal disease. Nat Med 24, 1342–1350 (2018). https://doi.org/10.1038/s41591-018-0107-6
19. Rachel Gordon, Robust artificial intelligence tools to predict future cancer MIT CSAIL, January 28, 2021, URL: https://news.mit.edu/2021/robust-artificial-intelligence-tools-predict-future-cancer-0128
20. Soliman A, Li Z, Parwani AV. Artificial intelligence's impact on breast cancer pathology: a literature review. Diagn Pathol. 2024 Feb 22;19(1):38. https://doi.org/10.1186/s13000-024-01453-w. PMID: 38388367; PMCID: PMC10882736.
21. Health tech start-up uses AI to redefine cardiology care, May 26, 2023, URL: https://www.gehealthcare.com/insights/article/health-tech-startup-uses-ai-to-redefine-cardiology-care
22. Lee EJ, Kim YH, Kim N, Kang DW. Deep into the Brain: Artificial Intelligence in Stroke Imaging. J Stroke. 2017 Sep;19(3):277–285. https://doi.org/10.5853/jos.2017.02054. Epub 2017 Sep 29. PMID: 29037014; PMCID: PMC5647643.
23. Johnson KB, Wei WQ, Weeraratne D, Frisse ME, Misulis K, Rhee K, Zhao J, Snowdon JL. Precision Medicine, AI, and the Future of Personalized Health Care. Clin Transl Sci. 2021 Jan;14(1):86–93. https://doi.org/10.1111/cts.12884. Epub 2020 Oct 12. PMID: 32961010; PMCID: PMC7877825.
24. Fernando Suarez Saiz et al., Artificial Intelligence Clinical Evidence Engine for Automatic Identification, Prioritization, and Extraction of Relevant Clinical Oncology Research. JCO Clin Cancer Inform 5, 102–111 (2021), https://doi.org/10.1200/CCI.20.00087
25. Zhixiang Wang, Artificial intelligence applications in oncology to augment data and support decisions, GROW – Basic and Translational Cancer Biology Radiotherapie, URL: https://cris.maastrichtuniversity.nl/en/publications/artificial-intelligence-applications-in-oncology-to-augment-data-, 11 Jul 2023, Maastricht University.
26. Daniel J. Mollura, MD, et al., Artificial Intelligence in Low- and Middle-Income Countries: Innovating Global Health Radiology, Radiology 2020; 00:1–8; https://doi.org/10.1148/radiol.2020201434
27. Acosta, J.N., Falcone, G.J., Rajpurkar, P. et al. Multimodal biomedical AI. Nat Med 28, 1773–1784 (2022). https://doi.org/10.1038/s41591-022-01981-2
28. Bohr A, Memarzadeh K. The rise of artificial intelligence in healthcare applications. Artificial Intelligence in Healthcare. 2020:25–60. https://doi.org/10.1016/B978-0-12-818438-7.00002-2. Epub 2020 Jun 26. PMCID: PMC7325854.
29. von Ende, E.; Ryan, S.; Crain, M.A.; Makary, M.S. Artificial Intelligence, Augmented Reality, and Virtual Reality Advances and Applications in Interventional Radiology. Diagnostics 2023, 13, 892. https://doi.org/10.3390/diagnostics13050892
30. Lu MT, Ivanov A, Mayrhofer T, Hosny A, Aerts HJWL, Hoffmann U. Deep Learning to Assess Long-term Mortality From Chest Radiographs. JAMA Netw Open. 2019;2(7):e197416. https://doi.org/10.1001/jamanetworkopen.2019.7416
31. Cui S, Traverso A, Niraula D, Zou J, Luo Y, Owen D, El Naqa I, Wei L. Interpretable artificial intelligence in radiology and radiation oncology. Br J Radiol. 2023 Oct;96(1150):20230142. https://doi.org/10.1259/bjr.20230142. Epub 2023 Jul 26. PMID: 37493248; PMCID: PMC10546466.
32. IBM Watson Health Introduces New Opportunities for Imaging AI Adoption, Nov 30, 2021, URL: https://newsroom.ibm.com/2021-11-30-IBM-Watson-Health-Introduces-New-Opportunities-for-Imaging-AI-Adoption
33. Sandhu SS, Gorji HT, Tavakolian P, Tavakolian K, Akhbardeh A. Medical Imaging Applications of Federated Learning. Diagnostics (Basel). 2023 Oct 6;13(19):3140. https://doi.org/10.3390/diagnostics13193140. PMID: 37835883; PMCID: PMC10572559.
34. Rieke, N., Hancox, J., Li, W. et al. The future of digital health with federated learning. npj Digit. Med. 3, 119 (2020). https://doi.org/10.1038/s41746-020-00323-1
35. Rehman MHU, Hugo Lopez Pinaya W, Nachev P, Teo JT, Ourselin S, Cardoso MJ. Federated learning for medical imaging radiology. Br J Radiol. 2023 Oct;96(1150):20220890. https://doi.org/10.1259/bjr.20220890. PMID: 38011227; PMCID: PMC10546441.

36. Darzi, E., Sijtsema, N.M. & van Ooijen, P.M.A. A comparative study of federated learning methods for COVID-19 detection. Sci Rep 14, 3944 (2024). https://doi.org/10.1038/s41598-024-54323-2
37. M. F. Sohan and A. Basalamah, "A Systematic Review on Federated Learning in Medical Image Analysis," in IEEE Access, vol. 11, pp. 28628–28644, 2023, https://doi.org/10.1109/ACCESS.2023.3260027.
38. John McCormack, Ethical Issues Loom as Artificial Intelligence Shows Promise for Health Information, February 23, 2024, URL: https://journal.ahima.org/page/ethical-issues-loom-as-artificial-intelligence-shows-promise-for-health-information
39. Sandhu SS, Gorji HT, Tavakolian P, Tavakolian K, Akhbardeh A. Medical Imaging Applications of Federated Learning. Diagnostics (Basel). 2023 Oct 6;13(19):3140. https://doi.org/10.3390/diagnostics13193140. PMID: 37835883; PMCID: PMC10572559.
40. Huang B, Yang F, Yin M, Mo X, Zhong C. A Review of Multimodal Medical Image Fusion Techniques. Comput Math Methods Med. 2020 Apr 23;2020:8279342. https://doi.org/10.1155/2020/8279342. PMID: 32377226; PMCID: PMC7195632.
41. Yen C, Lin CL, Chiang MC. Exploring the Frontiers of Neuroimaging: A Review of Recent Advances in Understanding Brain Functioning and Disorders. Life (Basel). 2023 Jun 29;13(7):1472. https://doi.org/10.3390/life13071472. PMID: 37511847; PMCID: PMC10381462.
42. Siemens Healthineers, Innovations with impact, URL: https://www.siemens-healthineers.com/innovations
43. Aman S Saini, Philips Incisive CT: A Breakthrough in Medical Imaging Technology, URL: https://www.linkedin.com/pulse/philips-incisive-ct-breakthrough-medical-imaging-aman-s-saini/?utm_source=rss&utm_campaign=articles_sitemaps&utm_medium=google_news, October 09, 2023.
44. Philips, Seven innovations in radiology workflow that are improving efficiency and quality of care, Nov 24, 2020, URL: https://www.philips.com/a-w/about/news/archive/blogs/innovation-matters/2020/20201124-seven-innovations-in-radiology-workflow-that-are-improving-efficiency-and-quality-of-care.html
45. Valentina Vitiello et al., Emerging Robotic Platforms for Minimally Invasive Surgery, IEEE Reviews In Biomedical Engineering, Vol. 6, 2013, URL: https://ieeexplore.ieee.org/stamp/stamp.jsp?arnumber=6392862
46. Thabit, A., Benmahdjoub, M., van Veelen, ML.C. et al. Augmented reality navigation for minimally invasive craniosynostosis surgery: a phantom study. Int J CARS 17, 1453–1460 (2022). https://doi.org/10.1007/s11548-022-02634-y
47. Palumbo A. Microsoft HoloLens 2 in Medical and Healthcare Context: State of the Art and Future Prospects. Sensors (Basel). 2022 Oct 11;22(20):7709. https://doi.org/10.3390/s22207709. PMID: 36298059; PMCID: PMC9611914.
48. de Farias, E.C., di Noia, C., Han, C. et al. Impact of GAN-based lesion-focused medical image super-resolution on the robustness of radiomic features. Sci Rep 11, 21361 (2021). https://doi.org/10.1038/s41598-021-00898-z
49. Skandarani, Y.; Jodoin, P.-M.; Lalande, A. GANs for Medical Image Synthesis: An Empirical Study. J. Imaging 2023, 9, 69. https://doi.org/10.3390/jimaging9030069
50. Akash Takyar, Generative Adversarial Networks (Gans): A Deep Dive Into The Architecture And Training Process, URL: https://www.leewayhertz.com/generative-adversarial-networks/
51. NVIDIA, Clara for Medical Imaging, URL: https://www.nvidia.com/en-in/clara/medical-imaging/
52. CloudMedx, Aggregation, Automation and AI, it's all in there, URL: https://cloudmedxhealth.com/
53. Radiology Partners Launches AI Integration Platform with AWS HealthImaging, URL: https://www.radpartners.com/2023/11/radiology-partners-launches-ai-integration-platform-with-aws-healthimaging/#:~:text=RP%20is%20now%20joining%20forces,across%20hospitals%20and%20health%20systems, November 27, 2023.
54. Nuance Accelerates the Adoption of Generative AI for Radiology with PowerScribe Advanced Auto Impression Capability, URL: https://news.nuance.com/2023-11-28-Nuance-

Accelerates-the-Adoption-of-Generative-AI-for-Radiology-with-PowerScribe-Advanced-Auto-Impression-Capability, Nov. 28, 2023.
55. Dias D, Paulo Silva Cunha J. Wearable Health Devices-Vital Sign Monitoring, Systems and Technologies. Sensors (Basel). 2018 Jul 25;18(8):2414. https://doi.org/10.3390/s18082414. PMID: 30044415; PMCID: PMC6111409.
56. Javad Pool, Saeed Akhlaghpour, Farhad Fatehi, Andrew Burton-Jones, A systematic analysis of failures in protecting personal health data: A scoping review, International Journal of Information Management, Volume 74, 2024, 102719, ISSN 0268-4012, https://doi.org/10.1016/j.ijinfomgt.2023.102719, (https://www.sciencedirect.com/science/article/pii/S0268401223001007)
57. Abouelmehdi, K., Beni-Hessane, A. & Khaloufi, H. Big healthcare data: preserving security and privacy. J Big Data 5, 1 (2018). https://doi.org/10.1186/s40537-017-0110-7
58. Huang J, Neill L, Wittbrodt M, et al. Generative Artificial Intelligence for Chest Radiograph Interpretation in the Emergency Department. JAMA Netw Open. 2023;6(10):e2336100. https://doi.org/10.1001/jamanetworkopen.2023.36100
59. Abedin, B., Meske, C., Junglas, I. et al. Designing and Managing Human-AI Interactions. Inf Syst Front 24, 691–697 (2022). https://doi.org/10.1007/s10796-022-10313-1
60. Brady, A.P., Allen, B., Chong, J. et al. Developing, purchasing, implementing and monitoring AI tools in radiology: practical considerations. A multi-society statement from the ACR, CAR, ESR, RANZCR & RSNA. Insights Imaging 15, 16 (2024). https://doi.org/10.1186/s13244-023-01541-3
61. Define-AI Directory, URL: https://www.acrdsi.org/DSI-Services/Define-AI
62. Morley J, Murphy L, Mishra A, Joshi I, Karpathakis K. Governing Data and Artificial Intelligence for Health Care: Developing an International Understanding. JMIR Form Res. 2022 Jan 31;6(1):e31623. https://doi.org/10.2196/31623. PMID: 35099403; PMCID: PMC8844981.
63. Mei, X., Lee, HC., Diao, Ky. et al. Artificial intelligence–enabled rapid diagnosis of patients with COVID-19. Nat Med 26, 1224–1228 (2020). https://doi.org/10.1038/s41591-020-0931-3
64. Yuxiang Liao, Hantao Liu, Irena Spasić, Deep learning approaches to automatic radiology report generation: A systematic review, Informatics in Medicine Unlocked, Volume 39, 2023, 101273, ISSN 2352-9148, https://doi.org/10.1016/j.imu.2023.101273, (https://www.sciencedirect.com/science/article/pii/S235291482300117X).
65. Thomas Dratsch et al., Automation Bias in Mammography: The Impact of Artificial Intelligence BI-RADS Suggestions on Reader Performance, May 2 2023, https://doi.org/10.1148/radiol.222176
66. Troya J, Fitting D, Brand M, Sudarevic B, Kather JN, Meining A, Hann A. The influence of computer-aided polyp detection systems on reaction time for polyp detection and eye gaze. Endoscopy. 2022 Oct;54(10):1009–1014. https://doi.org/10.1055/a-1770-7353. Epub 2022 Feb 14. PMID: 35158384; PMCID: PMC9500006.
67. FDA, Proposed Regulatory Framework for Modifications to Artificial Intelligence/Machine Learning (AI/ML)-Based Software as a Medical Device (SaMD) - Discussion Paper and Request for Feedback, April 2, 2019, URL: https://www.fda.gov/medical-devices/software-medical-device-samd/artificial-intelligence-and-machine-learning-software-medical-device
68. FDA, Safe Algorithmic Change Protocols for Modifications to AI/ML-based Software as a Medical Device, URL: https://www.fda.gov/science-research/advancing-regulatory-science/safe-algorithmic-change-protocols-modifications-aiml-based-software-medical-device
69. FDA, CDRH Issues Draft Guidance on Predetermined Change Control Plans for Artificial Intelligence/Machine Learning-Enabled Medical Devices, March 30, 2023, URL: https://www.fda.gov/medical-devices/medical-devices-news-and-events/cdrh-issues-draft-guidance-predetermined-change-control-plans-artificial-intelligencemachine
70. CDC, Health Insurance Portability and Accountability Act of 1996 (HIPAA), URL: https://www.cdc.gov/phlp/publications/topic/hipaa.html
71. What is GDPR, the EU's new data protection law?, URL: https://gdpr.eu/what-is-gdpr/

Appendices

Appendix 1

Exploring the diverse realm of medical imaging and analysis, this section delves into various applications and implementations of supervised machine learning techniques. From diagnosing diseases to enhancing image quality, these methods play a pivotal role in advancing healthcare through innovative technologies.

1. *Linear Regression*

   Linear regression is a statistical method used to model the relationship between a dependent variable and one or more independent variables [1]. The goal is to find the linear equation that best represents the relationship between these variables (Fig. A.1).

   In supervised learning, the algorithm is provided with input-output pairs, and its objective is to learn the mapping function that relates the inputs to the corresponding outputs. Linear regression fits well into this framework by establishing a linear relationship between the input features and the output variable, making it a valuable tool for prediction tasks.

Role in Medical Imaging

Linear regression can be employed for various purposes, such as predicting clinical outcomes, estimating disease progression, or analyzing the relationship between imaging features and patient characteristics.

Example: Estimating Tumor Growth

Consider a scenario with a dataset comprising medical imaging data of patients with brain tumors, encompassing details about tumor size (independent variable) and the corresponding patient's age (dependent variable). The application of linear regression to this dataset allows the algorithm to discern the linear relationship between tumor size and patient age. Following the model's training, it becomes capable of predicting potential tumor sizes based on the age of new patients.

E. Seeram, V. Kanade, *Artificial Intelligence in Medical Imaging Technology*,
https://doi.org/10.1007/978-3-031-64049-0

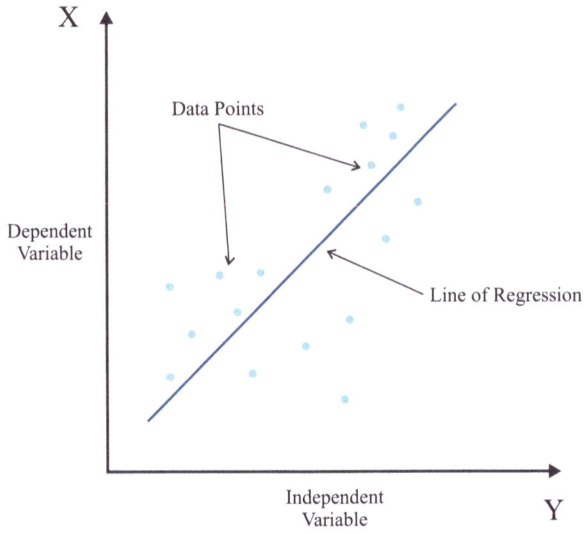

Fig. A.1 Linear regression. See text for further explanation

This application can assist medical professionals in understanding how patient age correlates with tumor growth, enabling them to anticipate potential developments and make informed decisions about treatment plans. Linear regression thus becomes a valuable tool in medical imaging, providing insights into the relationships between different variables and supporting data-driven decision-making in healthcare.

2. *Decision Trees*

A decision tree is a treelike model used for both classification and regression tasks in machine learning. It recursively splits the data into subsets based on the most significant feature at each node. These splits continue until a predefined stopping criterion is met, leading to terminal nodes or leaves where the final predictions or classifications are made [2] (Fig. A.2).

Decision trees are particularly attractive for their interpretability, as their structure mimics a series of humanlike decisions, making it easier to understand and visualize the decision-making process.

Role in Medical Imaging

Decision trees find valuable applications in medical imaging, offering interpretability and the ability to handle complex decision-making scenarios.

Example: Diagnosing Skin Lesions

Consider a scenario in dermatology where the task is to diagnose skin lesions as benign or malignant based on medical images. A decision tree can be constructed by considering various features such as color, size, and texture of the lesion. The algorithm learns to make decisions at each node based on these features, ultimately reaching a leaf node that represents the predicted diagnosis.

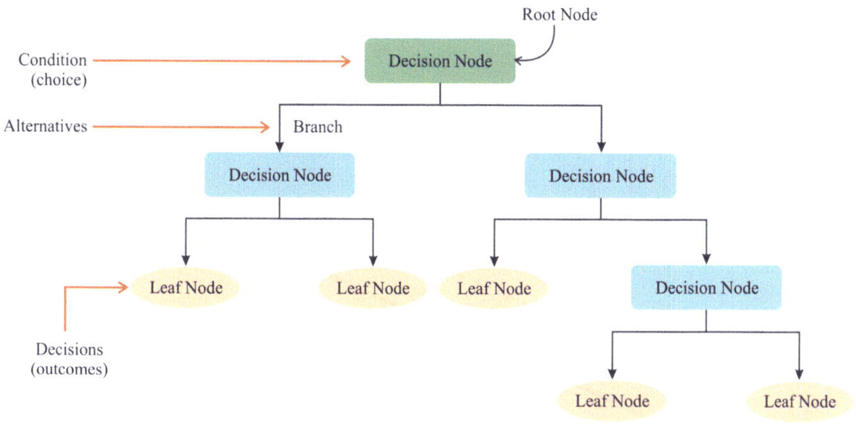

Fig. A.2 Decision trees. See text for further explanation

For instance, at the first decision node, the algorithm might assess whether the lesion's color is darker than a certain threshold. If yes, it may further split based on the lesion's size and so on. The final leaf nodes would provide predictions such as "benign" or "malignant" based on the learned decision rules.

With decision trees, clinicians can understand the reasoning behind a particular diagnosis. It also aids in identifying key features contributing to the decision, potentially offering insights into the characteristics of skin lesions associated with malignancy. Decision trees, therefore, serve as a valuable tool in medical image analysis, contributing to the development of effective and transparent diagnostic systems.

3. *Support Vector Machines (SVMs)*

Support vector machines (SVMs) are powerful supervised learning algorithms used for classification and regression tasks. SVMs aim to find a hyperplane in a high-dimensional space that best separates the data into different classes. The "support vectors" are the data points closest to the decision boundary, contributing significantly to defining the optimal hyperplane. SVMs are effective in handling both linear and non-linear relationships between features and labels by using various kernel functions [3].

The figure below illustrates the representation of support vectors (Fig. A.3):

Role in Medical Imaging

SVMs play a crucial role in tasks such as disease classification, image segmentation, and outcome prediction. The following example provides a detailed explanation on how SVMs function.

Example: Lung Nodule Detection in Chest CT Scans

Consider a dataset comprising features extracted from chest CT scans, focusing on characteristics such as nodule size, shape, and density. Each instance in the dataset is labeled as either benign or malignant based on expert annotations.

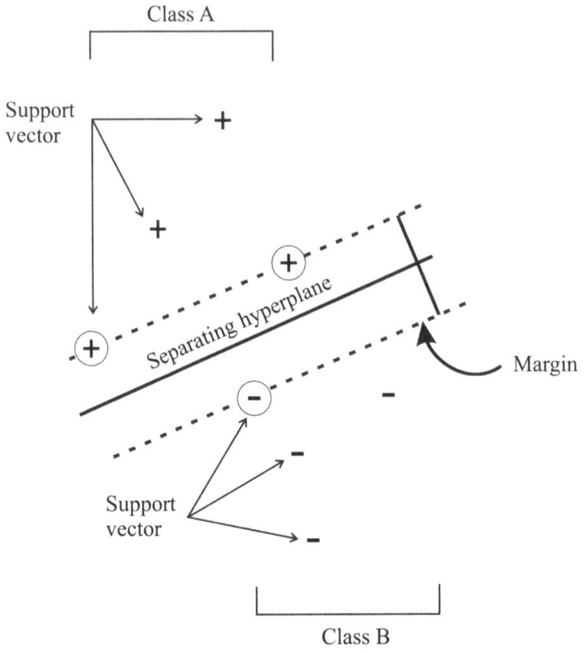

Fig. A.3 Support vectors. See text for further explanation

SVMs can be utilized to construct a classification model aimed at distinguishing between benign and malignant lung nodules. The SVM, in this context, determines the optimal hyperplane by considering crucial support vectors, defining the decision boundary that maximizes the margin between the two classes. By doing so, the SVM-based model becomes adept at accurately classifying new chest CT scans.

An SVM-driven model for lung nodule detection can be a valuable tool for doctors in the early identification of potential malignancies. Trained on a labeled dataset, the model can analyze new CT scans, providing predictions that assist radiologists in prioritizing cases and making well-informed decisions about patient care. This application of SVMs contributes to the progress of medical imaging, offering reliable automated diagnostic tools to support healthcare practitioners in their decision-making processes.

Appendix 2

The fundamental principles of two widely used clustering algorithms, namely, K-means and hierarchical clustering, will be explored to enhance comprehension.

1. *K-means Clustering*

K-means is a partitioning algorithm that aims to divide a dataset into a predefined number of clusters (K), where each data point belongs to the cluster with the nearest centroid [4]. Here's how K-means works:

- **Initialization**: It starts by randomly initializing K cluster centroids (mean points).
- **Assignment**: Data points are assigned to the cluster whose centroid is closest to them. This is usually based on a distance metric such as Euclidean distance.
- **Update**: The centroids of the clusters are recalculated as the mean of the data points within each cluster.
- **Iteration**: The assignment and update steps are repeated until convergence (i.e., centroids no longer change significantly) or for a fixed number of iterations.

In straightforward terms, K-means is a method for categorizing images into clusters based on their similarities. Consider a scenario where there is a collection of diverse fruits that need to be grouped by type. Applying K-means is much like arranging them into clusters of apples, bananas, and oranges. The algorithm examines the features of each fruit, such as shape and color, and assigns them to the appropriate cluster (Fig. A.4).

Typically, K-means can help us group similar X-rays, MRIs, or CT scans. For example, it can group images of healthy hearts, images of hearts with mild issues, and images of hearts with severe problems. This makes it easier for doctors to study and diagnose different conditions.

2. *Hierarchical Clustering*

Hierarchical clustering offers an alternative approach to grouping images, resembling the creation of a family tree. Consider a situation where there is a large family, and the goal is to comprehend the relationships among its members. Hierarchical clustering aids in organizing family members into branches, initiating with the most extensive groups (such as "relatives with heart issues") and subsequently subdividing them into more specific groups (such as "relatives with specific heart conditions").

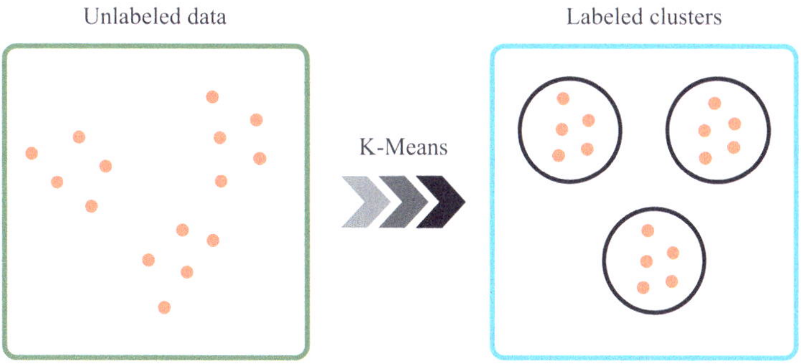

Fig. A.4 K-means clustering. See text for further explanation

This technique is handy when understanding the relationships between different types of medical images is necessary. For instance, it can help identify if certain types of tumors have subtypes, potentially leading to more precise treatments and improving patient care.

Technically, hierarchical clustering creates a treelike structure, or dendrogram, to represent the relationships between data points and clusters. It doesn't require specifying the number of clusters in advance [5]. There are two main approaches: agglomerative and divisive.

- **Agglomerative hierarchical clustering**: It starts with individual data points as separate clusters and then merges the closest clusters iteratively until a single cluster containing all data points is formed.
- **Divisive hierarchical clustering**: It begins with all data points in one cluster and recursively divides clusters until each data point is in its cluster.

Role in Medical Imaging

Hierarchical clustering is valuable in medical image analysis for various tasks:

- **Disease subtyping**: Hierarchical clustering can reveal natural groupings or subtypes within a patient population based on imaging features. This can aid in patient stratification for personalized treatments.
- **Image similarity**: It can be used to find images with similar characteristics, helping in image retrieval or the creation of image databases.
- **Segmentationvalidation**: It can assess the quality of image segmentation results by comparing them to ground truth segmentations.

Thus, unsupervised learning and clustering algorithms such as K-means and hierarchical clustering play a crucial role in discovering patterns and structures within medical imaging datasets. They play a role in tasks such as image segmentation, categorizing diseases, identifying anomalies, and various other applications, enhancing precision in diagnosis, treatment, and medical research.

Appendix 3

This section offers a summary and overview of various optimization algorithms that are well-suited for machine learning:

1. *Stochastic gradient descent* is an optimization algorithm widely used in machine learning particularly in the context of medical imaging and analysis. In this domain, SGD is employed to iteratively adjust the parameters of a neural network or other machine learning models to minimize the difference between predicted and actual outcomes. The "stochastic" aspect of SGD involves using randomly selected subsets of the training data (mini-batches) for each iteration, which introduces variability and helps prevent the algorithm from getting stuck

in local minima [6]. This is crucial for medical imaging tasks, as datasets can be vast and computationally intensive.

2. *Adaptive moment estimation (Adam)*: The Adam algorithm finds common application in machine learning. It combines ideas from both momentum optimization and RMSprop, adapting the learning rates for each parameter individually based on the historical gradients [7]. This adaptability makes Adam well-suited for non-stationary or noisy environments, which are common in medical imaging datasets.

Adam helps efficiently adjust the parameters of deep learning models during training to minimize the difference between predicted and actual outcomes. For instance, in the segmentation of medical images to identify organs or abnormalities, Adam's adaptive learning rates can be beneficial in navigating complex and variable image structures. This leads to quicker convergence and improved accuracy compared to traditional optimization algorithms.

3. *Root mean square propagation (RMSprop)*: RMSprop refers to an optimization algorithm that fine-tunes machine learning models. Its key role is in adjusting the learning rates for each parameter during training based on past gradients. In simpler terms, when teaching a computer to understand medical images, RMSprop helps the learning process by being sensitive to the importance of different features [8].

For instance, when delineating structures such as blood vessels in medical images, RMSprop ensures that the learning rates are adaptively adjusted to handle variations in image contrast and scale. This adaptability allows the algorithm to focus more on challenging areas where detailed adjustments are needed, contributing to more effective and precise outlining of structures in imaging applications.

4. *Adaptive gradient descent (AdaGrad):* AdaGrad is a type of optimization algorithm that trains machine learning models. Its distinctive feature lies in its adaptability to different parameters by individually adjusting the learning rates during the training process [8].

In the context of medical image analysis, where datasets often have diverse characteristics, AdaGrad shines by allocating higher learning rates to parameters with smaller gradients and vice versa. For instance, in the identification of specific anatomical landmarks, such as joints or bones, AdaGrad ensures that the learning process is finely tuned to pay more attention to subtle boundaries and intricate details in the image. This adaptability allows the algorithm to effectively navigate through variations in anatomical structures, contributing to improved accuracy in pinpointing and characterizing specific landmarks in medical imaging tasks.

5. *Limited-memory Broyden-Fletcher-Goldfarb-Shanno (LBFGS)*: LBFGS comes in handy when dealing with high-dimensional optimization problems. It belongs to the family of quasi-Newton methods and is well-suited for scenarios where memory efficiency is crucial.

For example, in characterizing tissue properties within magnetic resonance imaging (MRI) scans, LBFGS can be employed to adjust the model parameters for accurately capturing the nuanced features and patterns indicative of different tissue types. Its memory-based approach enables the algorithm to handle large datasets and complex image structures, contributing to improved characterization of tissue properties in medical imaging applications.

6. *Conjugate gradient: Conjugate gradient*, a helpful algorithm in medical imaging for fine-tuning machine learning models, excels in dealing with quadratic optimization challenges. Quadratic optimization involves finding the minimum or maximum of a quadratic function and helps improve the precision of models. Conjugate gradient efficiently navigates through the intricate parameter spaces of these quadratic optimization problems, making it valuable for enhancing the accuracy and effectiveness of machine learning applications in medical image analysis.

For instance, when fine-tuning the parameters of a model to highlight specific features, conjugate gradient iteratively adjusts these parameters, taking into account the unique characteristics of the image data. For example, in the delineation of neurological structures such as the brain's white matter from magnetic resonance images, conjugate gradient aids in optimizing the model parameters to precisely outline the intricate patterns and structures, contributing to improved accuracy in medical imaging tasks related to neuroimaging and brain analysis.

To summarize, SGD updates parameters with random data samples, Adam combines momentum and adaptive learning rates, while RMSprop adjusts learning rates based on recent squared gradients. AdaGrad tailors rates individually for each parameter, LBFGS is a memory-efficient optimization method, and conjugate gradient optimizes by minimizing conjugate directions.

Appendix 4

This section provides detailed explanations of several essential evaluation metrics for machine learning models.

1. **Classification Metrics**

In the field of medicine, machine learning models are frequently used for classification tasks, where the goal is to categorize data into different classes or categories. It's important for medical professionals to understand the various classification metrics to assess the performance of these models accurately.

Consider a scenario where a machine learning model is designed to analyze X-rays to identify issues such as tumors or abnormalities. Here are some ways to measure its performance:

- **Accuracy**: This is like checking how many times the model got it right out of all the cases it looked at. However, sometimes there are way more healthy cases than sick ones, so accuracy alone might not tell the whole story.

- **Precision**: Precision in this context pertains to the model's accuracy when identifying issues. For a radiologist, high precision implies that if the model indicates a problem, it is likely correct. This is vital to avoid unnecessary panic or treatments in medical scenarios.
- **Recall (or sensitivity)**: Now, this is about how good the model is at catching all the actual problems. High recall means the model is good at not missing any real issues, which is super important, especially in spotting diseases early.
- **Specificity**: This concept is akin to the opposite of recall. It focuses on the model's ability to accurately affirm, "Nope, everything looks good" when the situation is indeed favorable. This aspect is significant as it helps prevent unnecessary concern among patients.
- **F1-score**: The balance between precision and recall is essential. A machine learning model needs to be both careful and thorough, not just excelling in one aspect.
- **ROC curve and AUC**: Think of it like a summary graph. It shows how well the model can tell the difference between normal and problematic cases. A big area under the curve (AUC) means the model is good at this job.

For radiologists, comprehending these aspects contributes to enhanced trust and utilization of these models. They can assess the reliability of the model and determine when to place confidence in its findings. It is comparable to having a dependable assistant aiding in identifying potential issues in the images, thereby streamlining the process of diagnosing and treating patients with increased efficiency and accuracy.

2. **ROC Curve and AUC-ROC**

The receiver operating characteristic (ROC) curve serves as an assessment metric for binary classification challenges. It visually represents the trade-off between true positive rate (TPR) and false positive rate (FPR) across different threshold values, effectively distinguishing between the relevant and irrelevant outcomes. The area under the curve (AUC) quantifies a classifier's ability to differentiate between classes and acts as a condensed summary of the ROC curve.

The AUC-ROC curve acts as a performance gauge for classification tasks under various threshold configurations. The ROC curve, being a probability curve, finds expression in the AUC, signifying the model's capacity for separation. Essentially, it measures how proficiently the model can discern between different classes. A higher AUC implies enhanced predictive accuracy, effectively distinguishing between instances of 0 and 1. To draw an analogy, a higher AUC signifies the model's superior ability to differentiate between individuals with the ailment and those without.

Visualizing the ROC curve involves plotting TPR against FPR, with TPR on the y-axis and FPR on the x-axis. This graphical representation provides a comprehensive overview of the model's performance in navigating the sensitivity-specificity trade-off (Fig. A.5).

In the context of medical imaging, where the goal is often to distinguish between diseased and healthy tissues or conditions, ROC analysis becomes crucial.

Let's analyze the essential components of the ROC and AUC-ROC curve to grasp its significance better:

- **True positive rate (sensitivity)**: The proportion of actual positive cases correctly identified by the model.
  Importance: Sensitivity is vital because it measures the ability of the model to correctly identify patients with a particular condition. For instance, in cancer detection, sensitivity indicates how well the model can detect actual cases of cancer.
- **False positive rate (1—specificity)**: The proportion of actual negative cases incorrectly classified as positive by the model.
  Importance: Specificity is crucial to avoid unnecessary interventions or treatments for patients who do not have the condition. For example, in a mammogram, a false positive could lead to unnecessary biopsies.
- **ROC curve**: A graphical representation of the trade-off between sensitivity and specificity across different threshold values.
  Importance: The ROC curve provides a comprehensive overview of the model's performance at various discrimination thresholds. A curve closer to the upper-left corner indicates better overall performance.
- **Area under the ROC curve (AUC-ROC)**: A singular value encapsulating the model's overall performance, derived from the area under the ROC curve.
  Importance: AUC-ROC condenses the model's discriminatory prowess into a concise metric. A higher AUC-ROC translates to heightened ability in distinguishing between different medical conditions (Figs. A.6 and A.7).

Example

Consider a medical imaging scenario, specifically in the context of diagnosing lung nodules in chest X-rays. Radiologists employ a machine learning model designed to

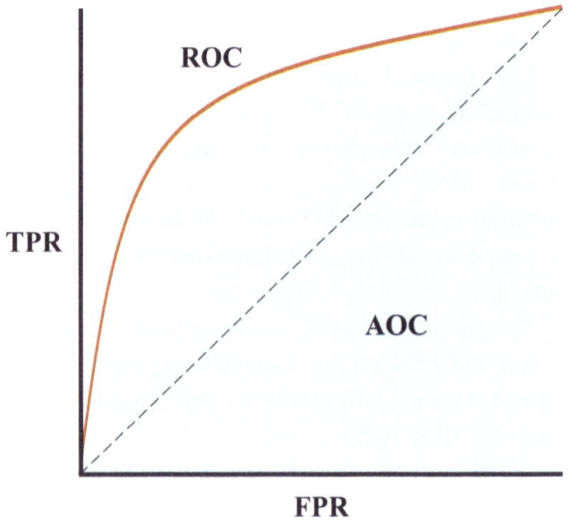

Fig. A.5 ROC curve and AUC-ROC. See text for further explanation

identify potential nodules, and the ROC curve and AUC-ROC become indispensable tools for assessing its performance.

The ROC curve visually illustrates the model's trade-off between sensitivity (detecting true positive cases, i.e., actual nodules) and specificity (avoiding false positives, such as benign structures mistaken for nodules) across varying decision thresholds. The AUC-ROC, representing the area under this curve, quantifies the model's overall ability to discriminate between images with actual nodules and those without. A higher AUC-ROC indicates superior diagnostic accuracy, empowering radiologists to gauge the model's proficiency and fine-tune its operational parameters based on the clinical context.

In the realm of medical imaging, the ROC curve and AUC-ROC offer radiologists a comprehensive evaluation of the nodule detection model's effectiveness. By visualizing the trade-off between sensitivity and specificity, radiologists can choose an optimal threshold that aligns with their priorities.

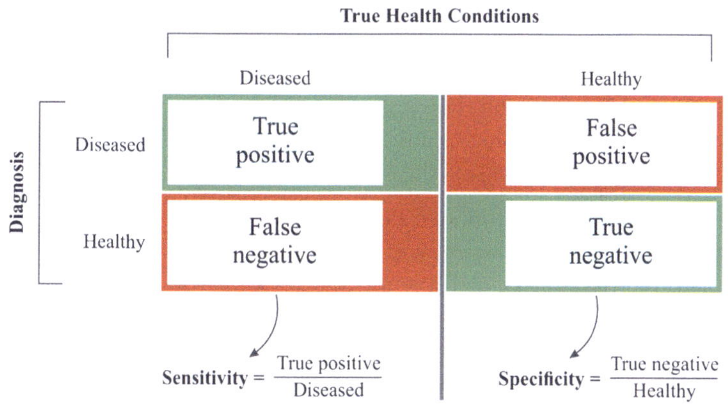

Fig. A.6 Determination of sensitivity and specificity values. See text for further explanation

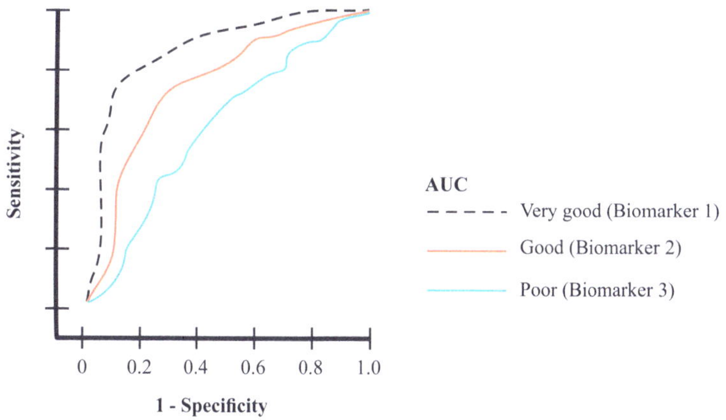

Fig. A.7 AUC-ROC. See text for further explanation

For instance, they may adjust the model to prioritize higher sensitivity, ensuring the identification of as many true positive cases as possible while still maintaining an acceptable level of specificity to minimize false positives.

3. **Precision-Recall Curve**

Precision and recall are two important metrics, especially in scenarios where class imbalances exist. In medical imaging, where the prevalence of certain conditions may be low, precision-recall analysis becomes crucial [9].

Imagine a chart that maps precision against recall—a tool particularly relevant in the medical field, especially when dealing with imbalanced data. This graph, known as the precision-recall curve, steps in as an alternative to ROC curves in scenarios where the data heavily leans toward one class. For medical professionals, this is particularly useful when evaluating a classifier's performance on detecting less prevalent conditions.

In the precision-recall curve, the aim is distinct from the ROC curve. Instead of striving for the upper-left-hand corner, as in ROC, the goal is to reach the upper-right-hand corner. This shift in focus is significant for medical practitioners, emphasizing the importance of achieving high precision in diagnoses while ensuring optimal recall, especially when dealing with conditions that are less common but demand accurate identification (Fig. A.8).

Grasping the key terminologies utilized in precision and recall metrics is essential to comprehend its relevance in the medical imaging.

- **Precision**: Precision is the ratio of true positive predictions to the total predicted positives (sum of true positives and false positives).

  Importance: Precision is vital when a false positive prediction can have significant consequences. For example, in the diagnosis of a rare disease, high precision ensures that the positive predictions made by the model are more likely to be correct.

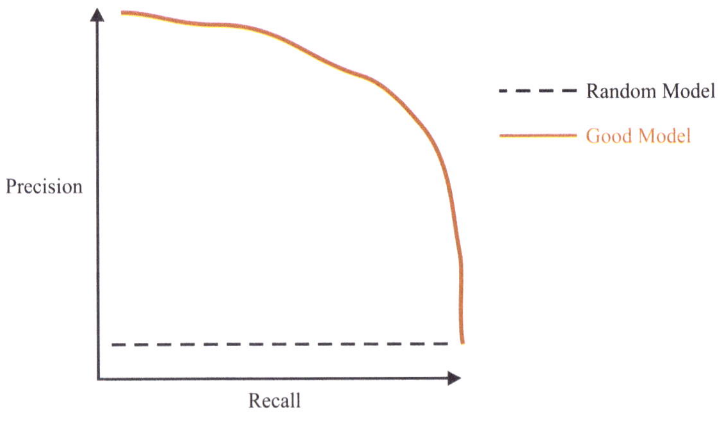

Fig. A.8 Precision-recall curve. See text for further explanation

- **Recall (sensitivity)**: Recall is the ratio of true positive predictions to the total actual positives (sum of true positives and false negatives).

  Importance: Recall is crucial when missing a positive case can have severe consequences. For instance, in cancer detection, high recall ensures that the model identifies most of the actual positive cases, minimizing false negatives.

- **Precision-recall curve**: A graphical representation of the trade-off between precision and recall for different threshold values.

  Importance: The precision-recall curve allows healthcare professionals to choose a threshold that balances the need for precision and recall based on the specific requirements of the medical task.

Example

Imagine a scenario where radiologists are using a smart tool to spot rare conditions, such as a specific type of heart problem in ultrasound images. This tool, powered by machine learning, gives them predictions, but it's crucial to know how reliable those predictions are. Here's where the precision-recall curve comes into play. Picture a graph that shows the trade-off between the tool's accuracy in spotting true positive cases (correctly identifying the heart issue) and its ability to catch all actual positive cases (recall).

This curve helps radiologists find the sweet spot where the tool is really good at getting things right (high precision) while also not missing out on any actual cases (high recall). It's like finding the right balance between being super accurate and not letting anything important slip through the cracks. This way, radiologists can adjust the tool to fit their priorities, making sure it's trustworthy and doesn't flag too many false alarms.

4. **Confusion Matrix Analysis**

A confusion matrix is a table that describes the model's performance by comparing the actual class labels with the predicted class labels. It provides the basis for calculating several classification metrics.

Key Components (Fig. A.9)

- True positives (TP): The number of correctly predicted positive instances.
- True negatives (TN): The number of correctly predicted negative instances.
- False positives (FP): The number of negative instances incorrectly predicted as positive (Type I error).
- False negatives (FN): The number of positive instances incorrectly predicted as negative (Type II error).

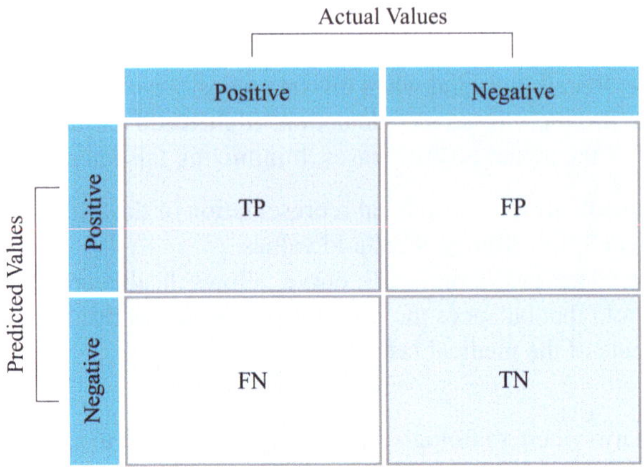

Fig. A.9 Key components of confusion matrix. See text for further explanation

Example

Consider a scenario where there is a binary classification model for breast cancer detection. The resulting confusion matrix would take the following form:

Actual/predicted	Positive (cancer)	Negative (no cancer)
Positive (cancer)	120 (TP)	10 (FN)
Negative (no cancer)	5 (FP)	865 (TN)

1. **Accuracy**

   Accuracy is a measure of the proportion of correct predictions made by the model. It's the ratio of correctly predicted instances (TP + TN) to the total instances.

Formula: Accuracy = (TP + TN)/(TP + TN + FP + FN)

Example

In the breast cancer detection example, the accuracy of the model would be (120 + 865)/(120 + 865 + 10 + 5) = 98.50%.

2. **Precision (Positive Predictive Value)**

   Precision measures the accuracy of positive predictions. It is the ratio of true positive predictions to the total predicted positive instances.

Formula: Precision = TP/(TP + FP)

Example

In breast cancer detection, precision measures how many of the predicted cancer cases are true positives. Precision = 120/(120 + 5) = 96%.

3. **Recall (Sensitivity, True Positive Rate)**

   Recall measures the ability of the model to correctly identify positive instances. It is the ratio of true positive predictions to the total actual positive instances.

Formula: Recall = TP/(TP + FN)

Example
In breast cancer detection, recall measures how many of the actual cancer cases are correctly predicted. Recall = 120/(120 + 10) = 92.31%.

4. **Specificity (True Negative Rate)**
   Specificity measures the ability of the model to correctly identify negative instances. It is the ratio of true negative predictions to the total actual negative instances.

Formula: Specificity = TN/(TN + FP)

Example
In breast cancer detection, specificity measures how many of the actual non-cancer cases are correctly predicted. Specificity = 865/(865 + 5) = 99.42%.

5. **F1-Score**
   The F1-score is the harmonic mean of precision and recall. It provides a balance between precision and recall. A higher F1-score indicates a better trade-off between precision and recall.

Formula: F1-Score = 2 * (Precision * Recall)/(Precision + Recall)

Example
In breast cancer detection, the F1-score would be calculated as 2 * (0.9524 * 0.9231)/(0.9524 + 0.9231) = 0.9373.

5. **Intersection over Union (IoU)andDice Coefficient**
   Intersection over union (IoU) and the dice coefficient are important evaluation metrics, particularly for tasks such as image segmentation, where the goal is to identify and outline regions of interest within medical images. These metrics are essential for assessing the accuracy of machine learning models in tasks such as tumor segmentation in radiology or cell segmentation in histopathology [10, 11]. The following section delves into the details of IoU and dice coefficient.

A. **Intersection over Union**
   Intersection over union, also known as the Jaccard index, is a metric used to assess the accuracy of object detection or image segmentation tasks. It measures the extent of overlap between the predicted and actual regions of interest. IoU is calculated as the ratio of the area of intersection between the predicted and actual regions to the area of their union.

Formula: IoU = (Area of Intersection)/(Area of Union)

Example
Consider a scenario where a medical image involves a machine learning model segmenting a region of interest, such as a tumor. The radiologist manually outlines the actual tumor region, and the IoU is employed to assess how well the model's segmentation aligns with the ground truth, represented by the radiologist's outline.

- If the area of intersection closely matches the area of the tumor outlined by the radiologist, and there is minimal or no extra region predicted by the model (the area of the union is close to the area of the actual tumor), the IoU will be close to 1, indicating excellent segmentation accuracy.
- Conversely, if the model's predicted region differs significantly from the actual tumor outline, or it predicts a much larger area (resulting in a large union area), the IoU will be lower, indicating poor segmentation accuracy.

IoU Usage

Medical imaging professionals use IoU when they need to be really precise about capturing the edges of larger things such as organs, lesions, or abnormalities in images. IoU is like a spotlight, focusing more on getting the boundaries of bigger structures right. If they want to be strict in checking how well their tool is outlining these larger areas, IoU is the go-to metric. It helps them make sure they're spot-on in marking the important parts in the images, especially when it comes to larger structures.

B. **Dice Coefficient**

The dice coefficient, also known as the Sørensen-Dice index, is another metric used for measuring the similarity between two sets. In the context of image segmentation, it assesses how well the predicted region aligns with the ground truth. The dice coefficient is calculated as twice the size of the intersection divided by the sum of the sizes of the two sets.

Formula: Dice Coefficient = (2 * Area of Intersection)/(Area of Predicted Region + Area of Actual Region)

Example

In a similar scenario as the one described for the IoU, the dice coefficient can be applied to evaluate the segmentation accuracy of the machine learning model.

- A high dice coefficient value, close to 1, indicates a strong overlap between the predicted region and the actual region, signifying accurate segmentation.
- A low dice coefficient value suggests limited overlap or significant differences between the predicted and actual regions, indicating poor segmentation accuracy.

Dice Coefficient Usage

Medical imaging professionals choose the dice coefficient when they need a fair and balanced look at the whole picture. It's like a friendlier metric that doesn't get too picky about the size of things in the images. So, if they're working on a task where structures come in all sizes, and they want to be good at catching both big and small things without being too strict, they lean toward the dice coefficient. It helps them find a nice balance between saying "yes, this is there" and "no, this isn't there" in a way that works well for structures of different sizes in medical images.

In summary, IoU and dice coefficient are valuable metrics that provide quantifiable measures of the alignment between predicted and actual regions of interest, helping medical professionals assess the model's performance and make decisions based on segmentation accuracy.

References

1. Kanade V. A., What Is Linear Regression? Types, Equation, Examples, and Best Practices for 2022, URL: https://www.spiceworks.com/tech/artificial-intelligence/articles/what-is-linear-regression/#:~:text=Linear%20regression%20is%20a%20statistical,Vijay%20Kanade%20AI%20Researcher
2. Berrar, D., Dubitzky, W. (2013). Decision Tree. In: Dubitzky, W., Wolkenhauer, O., Cho, KH., Yokota, H. (eds) Encyclopedia of Systems Biology. Springer, New York, NY. https://doi.org/10.1007/978-1-4419-9863-7_611
3. Zhang, X. (2011). Support Vector Machines. In: Sammut, C., Webb, G.I. (eds) Encyclopedia of Machine Learning. Springer, Boston, MA. https://doi.org/10.1007/978-0-387-30164-8_804
4. Kanade V.A., Top 10 Machine Learning Algorithms in 2022, URL: https://www.spiceworks.com/tech/artificial-intelligence/articles/top-ml-algorithms/
5. Halkidi, M. (2018). Hierarchical Clustering. In: Liu, L., Özsu, M.T. (eds) Encyclopedia of Database Systems. Springer, New York, NY. https://doi.org/10.1007/978-1-4614-8265-9_604
6. Bottou, L. (2012). Stochastic Gradient Descent Tricks. In: Montavon, G., Orr, G.B., Müller, KR. (eds) Neural Networks: Tricks of the Trade. Lecture Notes in Computer Science, vol 7700. Springer, Berlin, Heidelberg. https://doi.org/10.1007/978-3-642-35289-8_25
7. Mehdi Ghasri (2024). Adaptive moment estimation (Adam) (https://www.mathworks.com/matlabcentral/fileexchange/136679-adaptive-moment-estimation-adam), MATLAB Central File Exchange. Retrieved February 17, 2024.
8. Hassan, E., Shams, M.Y., Hikal, N.A. et al. The effect of choosing optimizer algorithms to improve computer vision tasks: a comparative study. Multimed Tools Appl 82, 16591–16633 (2023). https://doi.org/10.1007/s11042-022-13820-0
9. Kelly, B.S., Mathur, P., Plesniar, J. et al. Using deep learning–derived image features in radiologic time series to make personalised predictions: proof of concept in colonic transit data. Eur Radiol 33, 8376–8386 (2023). https://doi.org/10.1007/s00330-023-09769-9
10. Hesamian, M.H., Jia, W., He, X. et al. Deep Learning Techniques for Medical Image Segmentation: Achievements and Challenges. J Digit Imaging 32, 582–596 (2019). https://doi.org/10.1007/s10278-019-00227-x
11. Rashmi, R., Prasad, K. & Udupa, C.B.K. Breast histopathological image analysis using image processing techniques for diagnostic purposes: A methodological review. J Med Syst 46, 7 (2022). https://doi.org/10.1007/s10916-021-01786-9

E. Seeram, V. Kanade, *Artificial Intelligence in Medical Imaging Technology*, https://doi.org/10.1007/978-3-031-64049-0

Index

E. Seeram, V. Kanade, *Artificial Intelligence in Medical Imaging Technology*, https://doi.org/10.1007/978-3-031-64049-0

The manufacturer's authorised representative in the EU is Springer Nature Customer Service Centre GmbH, Europaplatz 3, 69115 Heidelberg, Germany. If you have any concerns regarding our products, please contact ProductSafety@springernature.com

Printed and bound by CPI Group (UK) Ltd, Croydon, CR0 4YY
30/09/2026
02225564-0004